T0179716

ANNALS *of* THE NEW YORK ACADEMY OF SCIENCES

EDITOR-IN-CHIEF
Douglas Braaten

ASSOCIATE EDITOR
Rebecca E. Cooney

PROJECT MANAGER
Steven E. Bohall

EDITORIAL ADMINISTRATOR
Daniel J. Becker

Artwork and design by Ash Ayman Shairzay

The New York Academy of Sciences
7 World Trade Center
250 Greenwich Street, 40th Floor
New York, NY 10007-2157

annals@nyas.org
www.nyas.org/annals

**The New York
Academy of Sciences**

Published by Blackwell Publishing
On behalf of the New York Academy of Sciences

Boston, Massachusetts
2011

ANNALS *of* THE NEW YORK ACADEMY OF SCIENCES

VOLUME
1236

ISSUE

Annals Meeting Reports

TABLE OF CONTENTS

Become a Member Today of the New York Academy of Sciences

The New York Academy of Sciences is dedicated to identifying the next frontiers in science and catalyzing key breakthroughs. As has been the case for 200 years, many of the leading scientific minds of our time rely on the Academy for key meetings and publications that serve as the crucial forum for a global community dedicated to scientific innovation.

 Select one FREE *Annals* volume and up to five volumes for only $40 each.

 Network and exchange ideas with the leaders of academia and industry.

 Broaden your knowledge across many disciplines.

 Gain access to exclusive online content.

Join Online at **www.nyas.org**

Or by phone at **800.344.6902** (516.576.2270 if outside the U.S.).

Ann. N.Y. Acad. Sci. ISSN 0077-8923

ANNALS OF THE NEW YORK ACADEMY OF SCIENCES

Issue: *Annals Meeting Reports*

Toward an interdisciplinary science of consumption

Stephanie D. Preston

Department of Psychology, University of Michigan, Ann Arbor, Michigan

Address for correspondence: Stephanie D. Preston, University of Michigan, Department of Psychology, 530 Church Street, Ann Arbor, MI 48109. prestos@umich.edu

Scientific perspectives on the drive to consume were presented in Ann Arbor, Michigan, at the conference entitled "The Interdisciplinary Science of Consumption: Mechanisms of Allocating Resources Across Disciplines." The meeting, which took place May 12–15, 2010 and was sponsored by Rackham Graduate School and the Department of Psychology at the University of Michigan, included presentations on human, primate, and rodent models and spanned multiple domains of consumption, including reward seeking, delay discounting, food-sharing reciprocity, and the consumption and display of material possessions across the life span.

Keywords: consumption; resource allocation; neureconomics; hoarding; decision making

Introduction

In a single half-hour of news, one can view segments about the rising rates of obesity, the devastating failure of our financial markets, and a woman who hoards. All of these stories have one thing in common: they are effects of *maladaptive* resource-allocation decisions.

Food, money, and material goods are necessary for survival. We have highly evolved and conserved mechanisms for making such important decisions—a capacity that is highly similar and perhaps homologous across mammalian species. For example, the same brain regions in the mesolimbocortical system are active in a diverse array of consumptive environments, including during single-cell firing in rats and monkeys to food rewards,[1] lesion studies of animal food hoarding,[2,3] hoarding in frontal patients,[4] functional imaging of people gambling or shopping,[5,6] and PET imaging of compulsive hoarders.[7]

However, this very makeup, which may have been adaptive in ancestral environments, also predisposes us toward the negative consequences of consumption, as we fall victim to the temporary and immediate rewards that abound in our society, such as unhealthy prepared foods, credit cards with high interest rates, and material luxuries that we cannot afford and are not environmentally sustainable. In this environment of plenty, we struggle to maintain a balance. And the problem is likely to get worse before it gets better. For example, consumable products that are rewarding are readily available, offer quick solutions to common problems, and are marketed in ever-sophisticated ways to tap our evolved propensities toward short-term reward (e.g., "no money down" programs with exorbitant interest rates for new furniture and plasma televisions). Second, the consequences of consumptive behavior are often hard to perceive because they are temporally and physically distant,[8] taking place in an unforeseen future or a distant location, such as in poorly regulated factories that maintain poor working conditions for laborers and destroy natural resources. The Nobel Prize–winning efforts of Al Gore and the current financial crisis have brought the issue of maladaptive consumption to the forefront of national politics. Academia needs to rise to the occasion to confront this issue and to develop unified models that can directly address the multiple negative consequences of this applied, social problem.

Given the significant problems in our society associated with consumption across domains (food, money, material goods), the devastating effects on health, finance, and the environment, and the preponderance of existing basic science knowledge, the time is ripe to form an interdisciplinary community

doi: 10.1111/j.1749-6632.2011.06163.x

interested in the study of consumption. To facilitate cross-disciplinary interaction and to search for common themes across domains, the term *consumption* is broadly construed as any process by which resources are acquired and/or processed, whether they are truly ingestible or only conceptually consumed, like material goods.

Many fields are already investigating pieces of this puzzle, but this is usually accomplished without any knowledge or direct study into the commonalities across species or domains. For example, marketing researchers study behavioral and neural mechanisms of choice using experimental designs almost identical to those used by other researchers to study pathological human compulsive hoarding. Economists have long studied resource-allocation decisions, such as tradeoffs between short- and long-term gains (intertemporal discounting), giving to one's self versus another (trust games), or between one's self and the community (the commons problem), and they have directly studied the power of possession upon people's psychology and behavior (the endowment effect). All of these are highly relevant to consumption but are typically not directly associated or studied in the context of similar phenomena in the food hoarding of animals, human pathological compulsive hoarding, or public policy issues of consumption and the environment. Research in neuroscience, neuroeconomics, and decision theory is currently focused on identifying the neural substrates of reward, decision making, and utility calculation—studies that are highly influential, often cited, and find consistent neural regions implicated in the processing and decisions about rewards (e.g., frontal and ventral striatal regions). However, there are other reasons for acquiring and saving that are more psychological—even existential—than have yet to be considered in these fields. For example, like the bowerbird, which decorates his nest with feathers and shards of glass to attract mates, people acquire objects to display wealth and their unique character (and often for the purpose of attracting a mate!). Some subjects even explicitly report *status* as a reason for possessing a prized object.[9] Similarly, the strongest motivations to keep possessions are stimulated by things that remind people of their past, a former self, or a social relationship.[10,11] However, these aspects of consumption are rarely discussed or rigorously tested by economists and neuroscientists, and even

within marketing and psychology, when they are discussed, they are not integrated conceptually with issues in animal behavior or pathological human behavior.

As evidence for the lack of centralized research on consumption, there is no term in psychology that refers to decisions about resources, and there is almost no psychological/neuroscientific research on decisions to acquire or discard material goods. *Acquisitiveness* is a term used by psychologists, but a search of this term in PsycInfo[a] reveals only 71 articles since 1887, only 13 of which are particularly germane to the topic, and most of which were written in the early 1900s (e.g., Cameron[12]). The term *hoarding* is used by animal and human researchers, but with very different connotations. *Resource-allocation decisions* is a term used by economists, but usually only for a specific scenario in which people keep money or give it to a stranger. There is a field called "material culture" (aka "material goods"), but articles on this topic are mainly restricted to anthropological and sociological treatments (a search of the term *psychology* in the *Journal of Material Culture* resulted in 40 articles since the journal's inception in 1996).[b]

To begin to address this gap in the academic literature, a conference entitled "The Interdisciplinary Science of Consumption: Mechanisms of Allocating Resources Across Disciplines" was held on May 12–15, 2010 at the Rackham Building on the University of Michigan central campus. The conference focused on mechanisms of resource-allocation decisions, such as acquiring and discarding important resources (e.g., money, food, material goods), with a specific goal to examine whether there really are deep commonalities in the mechanisms for such decisions across fields, species, and domains. The speakers were preeminent researchers from across fields, including marketing, economics, neuroscience, psychology, public policy, neuroethology, and animal behavior. The speakers were selected because their work informs the proposed unifying neurobiological and psychological model in which different forms of consumables or resources are assumed to be processed through largely shared

---

[a]http://www.apa.org/pubs/databases/psycinfo/index.aspx
[b]All data from June 2011.

neurobiological systems. The neurobiological systems were assumed to be those that evolved to motivate animals toward natural rewards that fulfill homeostatic needs, such as food and mates, but that could later be activated by artificial rewards and items that cannot be directly consumed, such as drugs and goods. More than a hundred people attended the academic portions of the meeting, with substantially more at the public lectures, producing a highly engaged audience of faculty and graduate students that attended all sessions together, discussed the themes over meals, and developed a narrative of discussion points throughout the meeting. In this way, the conference was like a large version of a "working meeting," where all attendees focused on a common goal of applying their expertise to a single, unified model of resource decisions. Below, we provide more detailed information on important components and themes of the meeting, organized by the order in which invited speakers presented their own work, followed by appendices that list members of the organizing committee and sources of funding for the meeting.

The allocation of food and aid in primates

The conference opened with an evening public lecture by Frans de Waal, a primatologist from Emory University and the director of the Living Links Center, who was introduced by his colleague, John Mitani. De Waal reviewed the evidence from his own laboratory research and from the anecdotal and experimental evidence across mammalian species for a sense of fairness, cooperation, and altruistic aid.[13] This work demonstrates that social mammals such as primates, elephants, and dolphins understand when another is in pain or need, and often act spontaneously to help, even in the absence of any immediate rewards to the giver. In addition, primates in particular have been shown to possess a sense of fairness and reciprocity, sharing food with those who have shared with them in the past, previously groomed them, or with whom they share a bonded, positive relationship (Fig. 1). These situational factors indicate that the animals utilize affective feeling states associated with the current need situation or the potential partner to make decisions about

Figure 1. Chimpanzee food sharing. Food sharing in chimpanzees is typically a peaceful affair in which individuals share, particularly with those with whom they share a bonded relationship (genetic or friendship) or to reciprocate for prior food sharing or grooming. In this photograph from the Yerkes Field Station, the female at the top right possesses the leafy branch as the female in the lower left reaches tentatively for a share. Photograph taken by Frans B. M. de Waal and reproduced with permission from Royal Society Publishing, Figure 2 (p. 2715) in Ref. 14.

giving. This work was a particularly useful start to the conference, because it established a key theme to which we would return throughout: evolved mechanisms for making decisions about resources critically rely upon social-affective states. Such states not only inform individuals of problems that need to be addressed, but also inform the decider about the level of security in the environment and the perceived utility of targets of consumption to which we may be driven. Many of the subsequent talks provided independent evidence that emotional security, particularly that which is established in the early home environment or social group, is a significant mediator of the type or level of consumption that an individual selects.

The neural bases of decisions to obtain rewards

In the first of four sessions, we explored the neural bases of decisions to obtain goods that give rewards. Antoine Bechara (professor of psychology and neuroscience at the University of Southern California and professor of psychiatry at McGill University) spoke first about the role of the insula (a brain region that represents internal, somatic, and affective states) in human addiction, particularly for smoking. Bechara used a two-system view to argue that most research on addiction focuses upon the role of an impulsive, automatic system involving the amygdala and striatum or a more reflective system in the prefrontal cortex that controls urges and makes planned, reasoned decisions. However, Bechara pointed to the insula as a region that can maintain poor impulse control in the face of addictive substances by conveying affective signals of craving that override the rational system and its ability to control behavior.[15] With Naqvi et al., Bechara studied patients who previously smoked and then quit without effort subsequent to a stroke that damaged their insula (Fig. 2).[16] For example, one man started smoking at the age of 14 and was up to 40 unfiltered cigarettes per day when he suffered a stroke at the age of 28, after which "[his] body forgot the urge to smoke." He simply no longer had the urge and did not have trouble resisting after that. Across 19 patients with damage to the insula and 50 with lesions not affecting the insula, 16 smokers quit without effort immediately after the damage, 12 of which were insula lesion patients. The majority of lesion patients

who smoked quit immediately, while the majority of noninsula lesion patients did not. Bechara suggests that the insular cortex (particularly the anterior insula) represents interoceptive signals that signal "urges," such as those that occur in response to cues of reward during homeostatic imbalance or deprivation, which in turn can "hijack" the more reflective control decision-making systems in the prefrontal cortex, biasing individuals toward the impulsive, unreflected, and powerful urge to consume characteristic of true addiction.[15]

This work is highly convergent with that of Terry Robinson (Elliot S. Valenstein Collegiate Professor of Behavioral Neuroscience and professor of psychology at the University of Michigan), who also presented work on the role of reward cues during addiction, which he studies through biopsychological animal models. Robinson used Pavolvian conditioning methods to demonstrate individual differences in the response to conditioned cues that a natural (unconditioned) reward is coming. In animals who are "sign tracking," a neutral stimulus that predicts a natural reward obtains an incentive salience that is inherently rewarding, motivating, and attended to; in contrast, in individuals who are "goal tracking," the conditioned stimulus predicts the reward but does not confer its own motivating incentive properties, allowing the animal to focus on the goal after delivery of the cue (Fig. 3).[17] Thus, for example, a sign-tracking rat who learns that an illuminated bar predicts the receipt of food through a chute to the left will attend to the bar during the delay, pressing it and investigating it while waiting for the food, whereas the goal tracker notes the light and then attends directly to the chute through which the food will be delivered.[18] These individual differences represent two bimodal peaks in the distribution and profoundly reflect the neurobiology of the individuals and their response to cues of reward. For example, sign trackers are more impulsive, more prone to explore a novel environment, reinstate drug seeking more quickly after extinction from the cue, but also reduce consumption in the absence of the cue.[19] These traits are associated with greater engagement of mesolimbic dopaminergic reward circuitry associated with the cue, greater sensitivity to dopamine agonists, greater dopamine release in the accumbens, and greater dopamine release in response to the cue after the association is learned, which does not occur in goal trackers.[20,21]

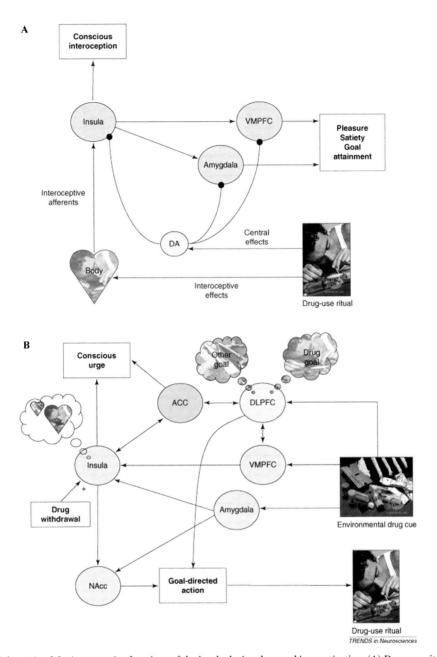

Figure 2. Schematic of the interoceptive functions of the insula during drug-seeking motivation. (A) Drug-use rituals produce interoceptive effects represented by the insula that produce the subjective qualities of the ritual, including conscious awareness of the drug's effects as well as the biological rewards (i.e., pleasure and satiety). Dopamine (DA) release from the drug's central effects may modulate the interoceptive rewards represented in the insula while causing the drug to become associated with pleasure, leading to future states of motivation and desire to obtain the drug. (B) Environmental cues associated with the drug, such as spatial contexts and drug paraphernalia, can reactivate the interoceptive representations in the insula via the VMPFC and amygdala, producing a subjective feeling of "craving" or a conscious urge to obtain the drug. This interoceptive representation feeds into the nucleus accumbens (NAcc), motivating actions toward the reward, while the dorsolateral prefrontal cortex (DLPFC) holds representations of the drug in mind and directs attention toward it, producing goal-directed actions toward the drug. The anterior cingulate cortex (ACC) additionally participates in the conscious feelings of urge through integrated representations of the insula's interoceptive states and the environmental cues of the drug, as well as by monitoring conflict associated with drug taking versus other goals. The insula may also mediate physiological signals associated with drug withdrawal that affect this process. Reprinted from Ref. 15, with permission from Elsevier.

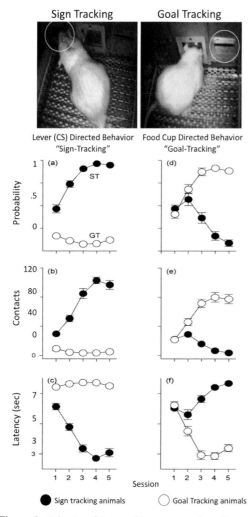

Figure 3. Behavior of sign-tracking versus goal-tracking rats during Pavlovian conditioning. The photograph at the top-left depicts a rat in a conditioning chamber in which the lever on the left illuminates (the conditioned cue) to predict the delivery of food in the door to the right (the unconditioned reward). The sign-tracking animals (depicted on the left of the photograph and with filled circles on the graphs) are more likely to interact with the lever in probability (A), number of contacts (B), and latency to contact (C), whereas the goal-tracking animals (depicted in the photograph on the right and with unfilled circles on the graphs) are more likely on each of these measures (D–F) to interact with the food cup to the right where the reward is actually delivered. Photograph reprinted from Ref. 17, with permission from Elsevier. Graphs reprinted from Ref. 18, with permission from Elsevier.

This work is critical because it shows a clear way in which individuals may differ in their perception of stimuli (drugs, people, or goods, for example) and the rewards they expect to receive from them. Sensitivity to cue-induced motivation to obtain rewards

can be highly explanatory of drug-addictive behavior and material consumption, as individuals are known to relapse in environments associated with prior drug taking and to show hedonic responses to cues like favored marketing brand images and even credit cards.

The conference was designed to investigate the role of such biological systems that clearly subserve the processing of natural, ingestible rewards, in the consumption of noningestible material rewards that can be purchased, hoarded, or displayed. Brian Knutson (associate professor of psychology and neuroscience, Stanford University) presented evidence for such crossover from his human neuroimaging work, finding that regions in the mesolimbocortical system produced unique signals of value and decision making while subjects made decisions to purchase material goods. With Rick et al.,[6] Knutson used his "SHOP" task to separately investigate the neural correlates associated with the passive viewing, pricing, and decision-making phases of material purchases (Fig. 4). He demonstrated that the nucleus accumbens (NAcc) is particularly involved during passive viewing of goods that people eventually purchase, implicating this region in processing the associated gains or rewards that motivate purchasing decisions. The insula responded particularly when subjects were presented with excessive prices and predicted subjects' susceptibility to the "endowment effect" (overvaluing products they own), perhaps reflecting a role for the insula in the anticipation of loss. Activity in the orbitofrontal cortex (OFC) declined with the insular response to excessive prices and pervaded during the decision phase, presumably because it integrates affective signals from regions like the NAcc and insula into an adaptive decision to obtain desired goods at their value. Thus, this work, along with that of Terry Robinson (see above), indicate that the same neural systems (particularly the dopaminergic NAcc and OFC) can mediate not only natural, unconditioned rewards like food or drugs, but also nonorganic, secondary, or learned rewards such as the cues that predict reward and material goods. Future work can investigate the origin of the rewards associated with material goods, dissociating perhaps signals associated with the predicted logistical benefits associated with an item (e.g., that addresses a current problem or need), the immediately available aesthetic qualities of the item (e.g., color, shininess, beauty),

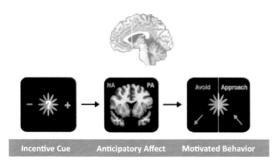

Figure 4. Knutson's anticipatory affect model. In the first phase, an incentive cue for an uncertain future outcome elicits activation in at least two brain regions (NAcc and anterior insula), which may correlate with the subsequent anticipatory affect response of either positive arousal (orange circles) or negative arousal (blue circles). In the final phase, the resulting overall signal (positive versus negative) then promotes the motivated behavior, which is either an approach (orange) or avoidance response (blue) to the cued outcome. For more information, see Ref. 22.

and the social rewards associated with obtaining a high-status item.

All three neuroscientific talks in the first session illuminated the general theme of the meeting, as they implicated overlapping neural systems between natural and material rewards, while providing novel details about the mechanism. In particular, the role of individual differences from Robinson's work in the response to cues of reward that maintain addictive behavior, as well as the role of the insula in Bechara and Knutson's work, represent mechanisms that are not common knowledge in the field, allowing us to form a more nuanced understanding of the underlying systems that can be exported to inform other domains.

The motivation to store versus use resources

In the second session, three presenters from different fields discussed how their work informs the development of the motivation to store versus to use resources. Bruce J. Ellis (professor of family studies and human development, John & Doris Norton Endowed Chair in Fathers, Parenting, and Families at the University of Arizona, Norton School of Family and Consumer Sciences) studies how human development, from a life history perspective, causes shifts in the bias toward risk-seeking behavior, such as drug abuse and teen pregnancy. Ellis' talk particularly emphasized how the transition into puberty

affects the reallocation of one's resources, with individuals transitioning from focusing their childhood investments into somatic problems like maintaining health, increasing physical growth, and developing sociocompetitive competencies into more reproductive efforts in the body and behavior. During puberty, males and females both become interested in sex and competing for sex, but through different qualitative approaches that are consistent with an evolutionary view. Males become more risk seeking by displaying their abilities and risks as a sexual display, while females engage in female–female competition for higher quality males and attempt to build and benefit from social coalitions. These propensities interact with the level of security versus stress in one's environment, with less safe or secure environments predisposing individuals toward particularly short-horizon strategies such as greater risk seeking through dangerous behaviors and status signaling through expensive consumables.[23] Just as Wilson and Daly[24] characterized these features in *young male syndrome*, Ellis points to a possible corollary syndrome in psychosocially stressed females, characterized by early sexual maturation, impulsive mate choice, low-quality parental investment, single motherhood, and earlier and more conspicuous consumption of sexualized products. This presentation outlined themes that were particularly necessary to pull together research across the meeting, as the role of the environment and its level of security, as well as the role of mate signaling through consumption, appeared to be key determinants of behavior in subsequent talks, but was not yet addressed in the more neuroscientific framework around which the conference was organized.

For example, David Sherry (professor of behavioral and cognitive neuroscience, Department of Psychology at the University of Western Ontario) subsequently presented his work on the proximate factors that motivate birds to store seeds for the winter. He provided an overview of research from optimality models in which cache decisions are influenced by environmental and internal variables, including the size of a food patch, the abundance and variability of food, the cost of carrying energy reserves as fat, the time of day, and the risk of starvation. His own research showed neural specialization in the hippocampus for remembering the location of stored food. The hippocampus

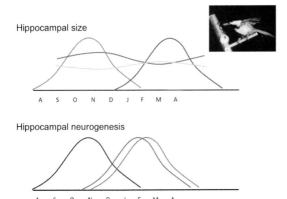

Hippocampal size

A   S   O   N   D   J   F   M   A

Hippocampal neurogenesis

A   S   O   N   D   J   F   M   A

Figure 5. Changes in hippocampal size and neurogenesis over the months of the year in food-storing birds. The image to the top right depicts a black-capped chickadee, a food-storing bird. Letters on the *x*-axis represent the months of the year, from August to April. The seasonal pattern in the hippocampal size and neurogenesis is variable, as shown in this schematic illustration that summarizes the results of a number of studies.[25] For hippocampal size, one study shows a fall peak, one shows a spring peak, and others show no systematic seasonal change at all. Seasonal variation in neurogenesis is more reliably observed, though one study reports a fall peak and two others report peaks in mid-winter and spring. Figure and image provided by David Sherry.

exhibits not only greater size in storing versus nonstoring bird species, but also plasticity, including neurogenesis, across seasons in which food storing is and is not a priority (Fig. 5).[25] Contrary to what one might expect, factors such as day length, which correlate highly with the seasons in which food storing take place, do not causally produce changes in brain morphology. Rather, the actual storing behavior appears to dynamically affect the recruitment of new neurons into regions associated with memory for the locations of caches.

Studies of animal food storing are important to demonstrate the capacity for an evolved neural system, even in a nonmammalian species with a small absolute brain size, to optimize decisions about resources. Animals clearly integrate environmental and internal body state conditions to adaptively improve their chances of survival and reproductive success. Humans also keep stores of many different kinds that they create and maintain in order to satisfy perceived future needs, including bank accounts, investments, food pantries, and emergency supplies (and perhaps even stores of other things

like social partners, memories, and ideas). Material goods appear to convey both the status and signaling functions discussed by Ellis, as well as the utilitarian future functions that are more associated with the storage of food in animals. For example, people keep books on the shelf both that reflect their identity and can be read in the future to abate boredom or to inform a particular problem. People also keep supplies for gift giving that can signal to someone that they care, to reciprocate a prior gift, and to signal their own wealth and largesse.

These fundamental resource decisions appear to be affected in compulsive hoarders, who acquire and fail to discard excessive numbers of items in their home to the point of distress, impaired daily living, and compromised safety. Randy Frost (Harold & Elsa Israel Professor, Department of Psychology, Smith College) is the world's leading expert on human compulsive hoarding, and his talk was attended by additional local public health workers and clinicians hoping to learn more about this disorder, which is very hard to treat. Frost discussed the diagnostic criteria for compulsive hoarding, which has largely been assumed heretofore to be a variant of obsessive compulsive disorder (OCD). However, Frost *et al.* are challenging this view, demonstrating that most hoarders do not have comorbid OCD, though some do, and many have comorbid depression, anxiety, and social phobia, which are not characteristic of OCD.[26–28] Additionally, SSRI medications that typically alleviate symptoms of OCD are less effective for compulsive hoarding, which responds better to cognitive-behavioral therapy (CBT) that is specifically designed to treat hoarding.[29] There were also significant themes of social and emotional trauma and anxiety in hoarders that resonated with the life-history strategy themes of Bruce Ellis. For example, hoarders experienced significant life-trauma events that appear to cause cognitions of worthlessness, mood dysregulation, and trouble inhibiting behavior, while they seek items that fulfill needs otherwise provided by bonded relationships with other people.[30] The role of sexual signaling through possessions in hoarders was discussed after the talk, because it played a prominent role in the life-history strategy approach espoused earlier. To date, there is no research specifically examining this, though a signaling motivation is consistent with many hoarding behaviors and should be investigated further.

The evolution and psychology of monetary saving and material consumption

Consumption as pollution: why other people's spending matters

Robert Frank (Henrietta Johnson Louis Professor of Management and professor of economics at Cornell University) is the author of *Luxury Fever*[31] and *Passions within Reason*,[32] and is now more recently known for his work on the role of emotion in conspicuous consumption in modern Western society. His evening public lecture was highly convergent with that of the previous and subsequent speakers, encapsulating the emergent theme in which social display and sexual selection are considered major contributors to the drive to consume. Frank presented a view of modern consumption as a social comparative ratchet that increases exponentially over time. According to this view, people are less concerned about their absolute level of wealth and material excesses than they are with looking good relative to their neighbors and peers. Thus, once new, larger, and more luxurious goods are introduced into the market (e.g., mega-homes), a precedent is set that must be reached and exceeded by competing peers, who must subsequently purchase even bigger homes, establishing a new precedent for the scale of luxury, and on *ad infinitum*. Recent economic practices, such as giving people loans that they cannot afford to repay or zero percent–down mortgages, augment this natural progression by introducing new standards of living that individuals fight to attain despite the fact that they do not have the means or financial security to afford them. Moreover, such excessive conspicuous consumption is not sustainable for the natural environment. Frank recommended a taxation approach in which individual taxes are exempt for savings but luxury purchases are taxed at a progressive rate. The meeting's social and evolutionary speakers were also focused on social and sexual display (i.e., performing actions for the purpose of attracting social partners and mates) as the driving motivation behind the evolution of material consumption and, thus, Frank's talk provided a welcome level of detail on this important topic that needed to be integrated with the existing biologically based framework.

The development and psychology of economics

On the second full day of the meeting, in the third session, Paul Webley (director and principal, School of Oriental and African Studies; professor of economic psychology, University of London; and visiting professor, School of Psychology, University of Exeter) and Stephen Lea (School of Psychology, University of Exeter) traveled from the United Kingdom to present their work on the development and psychology of economics. Lea and Webley are well known for their theoretical piece in *Behavioral and Brain Sciences* entitled, "Money as tool, money as drug: the biological psychology of a strong incentive."[33] They argued in this piece that people are motivated to obtain money both as a tool and as a drug, again asserting a cross-domain mechanism for natural and artificial/symbolic rewards. At the meeting, Webley first presented his many, ingenious behavioral experiments with children to discover their changing concepts of money, saving money, and the factors that predict successful saving. He found that young children (by six years of age) have the concept of saving money, which they already associate with patience, control, and the virtuous delay of gratification; however, they do not like the process. Concepts of saving and behavioral strategies for saving are fully functional by 12 years of age, though because of the continued dislike of the painful process, children creatively attempt to attain their consumptive desires without saving. In one study, mothers and grandmothers had more influence over children's saving habits, but most of the variance is explained by overall parenting style rather than by any particular saving practice or discussion with the child. For example, authoritative parents raised children who were more future-oriented in general, saved more, thought saving was a good practice, found it easier to save and to resist temptation, and used saving as a way to obtain money—these factors appeared to benefit from a mediated effect of increased conscientiousness and self-efficacy in saving and a reduced feeling that saving was a struggle. Children of "overinvolved" parents had similar qualities overall but were less likely to use saving as a way to obtain money, instead negotiating with parents to borrow or obtain more. Webley concluded that the development of saving requires learning about what is valued, learning strategies and habits for saving, and acquiring self-knowledge that is necessary to plan for delayed gratification.

Stephen Lea addressed the value of the concept of hyperbolic discounting to explain major life economic choices such as saving, planning for retirement, credit abuse, and consumer debt. He pointed out that the vast majority of research on discounting (impulsively preferring an earlier smaller reward to a larger, later reward), which was performed with rats and pigeons, uses experimental delays on the order of seconds and is probably mediated by largely automatic, unreflected cognitive processes. In contrast, major life financial decisions, such as planning for a vacation or paying off a credit card debt, take place over the course of months to decades and often involve substantial explicit thought and planning, even when decisions go awry. Lea argued that to explain such long-term financial acts, we need to augment concepts of temporal discounting with "mental time travel," which is a human cognitive process in which people recall past events and envisage future events in the service of the decision. However, the processes enabling mental time travel may not allow people to compare values across time accurately, producing biases that reflect both the effects of hyperbolic discounting as well as the errors from mental time travel (e.g., optimism and availability biases).

The final talk of the third session by the author of this article, Stephanie D. Preston (Department of Psychology, University of Michigan), presented the overview of the meeting theme by demonstrating how the neural substrates and psychological mechanisms for resource hoarding were shared across species and domains. Evidence from prior experiments was presented from food storing in rodents and human compulsive hoarding, shopping, and gambling, all of which implicated the mesolimbocortical system, particularly the NAcc and OFC. Data from a neuroimaging study in her own lab were presented[34] that demonstrate a role for the OFC across consumption decisions (e.g., acquiring or discarding goods, for personal use or monetary profit). The NAcc may only be involved to the extent that participants are acquisitive, which potentially suggests that hoarders are like the "sign-tracking" rats of Terry Robinson (above), as they appear to view the cues of reward as rewarding themselves. Additional work in her lab confirmed that acquisitive tendencies are normally distributed in the population (Fig. 6) and are particularly associated with underlying differences in anxiety,[35] which

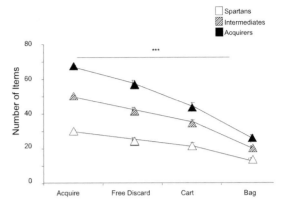

Figure 6. Normal, individual differences in acquisitiveness. Experimental data from the object decision task (ODT) divide subjects into three groups based on the number of objects that they acquire and keep across blocks of the task. Acquirers (those who take a lot but are not clinical hoarders) are presented in filled triangles, intermediates in hatched triangles, and Spartans in unfilled triangles. The blocks (on the *x*-axis) represent decisions to acquire as many hypothetical goods as one desires from a set shown one at a time, followed by an opportunity to cast off any acquired items that are no longer wanted, first without any pressure, and then after space constraints are introduced to fit items into a shopping cart and then a paper grocery bag.[35]

may adaptively provide individuals with resources that improve survival or reproductive success in environments perceived as uncertain or threatening.[36] However, the role of anxiety appears complex, with only a particular type of anxiety being involved that overlaps with, but is dissociable from, the anxiety of OCD.[26] Thus, the evidence again suggests that feelings of safety and security are critical drivers in the desire to consume; in addition, such drivers may directly enhance the perception of goods that satisfy such motivations.

Psychological and evolved underpinnings of the drive to consume

In the fourth session, human behavioral researchers presented information on the psychological and evolved mechanisms of risk preferences and the drive to consume. Kathleen Vohs (associate professor of marketing, McKnight Presidential Fellow, and Land O' Lakes Professor for Excellence in Marketing, Carlson School of Management, University of Minnesota) is a prolific social psychologist working in a department of marketing to examine the role of self-control to explain many real-world successes and failures to engage in desirable consumptive behavior (such as saving instead of spending, and

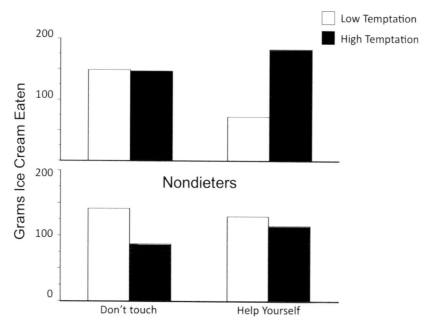

Figure 7. Chronic dieters: effects of self-regulatory depletion on ice cream consumption. Under conditions of high temptation, chronic dieters are particularly prone to eat ice cream when they can help themselves. Modified after Vohs and Heatherton.[37]

withholding instead of bingeing). Vohs described her model in which self-regulation is like a muscle that can be depleted and built up slowly overtime. This control muscle reflects a general, limited resource that can be tapped to control impulses and desires but that also can be depleted by one task and then be deficient to permit self-regulation in a subsequent, unrelated task. Vohs demonstrated such effects across domains, including dieting, impulsive spending,[38] and interpersonal behaviors with romantic partners, friends, and strangers.[39] Across domains, the worst effects of depletion—for example, from a boring, frustrating, or challenging task—are achieved by those trying to control their subsequent behavior, because chronic exertion in their domain of interest (e.g., dieting) leaves them vulnerable to failures in will power (Fig. 7).[40] She similarly demonstrated the need for self-resources in decision making[41] and that such self-regulation makes time move more slowly,[42] and once depleted, people think less rationally.[40,43] She also suggested that self-affirmation may be a beneficial, efficient way to reduce the effects of depletion on performance.[44] As in the prior talks, the consistency with which one's mental resources affect decisions across domains (financial, social, dietary) suggests that a common underlying biological system is implicated across

resource-allocation decisions. In addition, her work on the more controlled, rational cognitive system, which was alluded to by Bechara (above), helps to fill gaps in our understanding of how such control systems can actually permit successful control but also be impaired during decisions to consume.

Geoffrey Miller (associate professor, Department of Psychology, University of New Mexico) is an evolutionary psychologist who, like Robert Frank, emphasized the role of sexual selection in driving conspicuous consumption, popularized in his mainstream nonfiction book, *Spent: Sex, Evolution, and Consumer Behavior.*[45] In a collaboration with Griskevicius *et al.*,[46] Miller *et al.* found that romantic primes (e.g., attractive opposite-sex faces and writing about an ideal date), compared to control prime images, caused men, in particular, to increase spending for conspicuous consumption items that signal status and wealth (e.g., a new car, watch, holiday), while females increased spending on inconspicuous consumption items (e.g., medicine and household appliances) and were more likely to report giving in volunteer opportunities that would be publicly observed (Fig. 8). Offers of private help were more common in women than men but were not affected by romantic motives. Thus, males appear more likely to use status goods to signal their

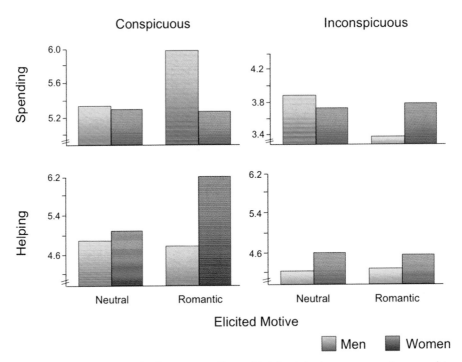

Figure 8. Effects of romantic or control primes on spending and helping behavior across conspicuous and inconspicuous conditions for males and females. Males (blue bars), in particular, increase conspicuous spending after romantic priming, whereas females (red bars) increase conspicuous helping and spending for inconspicuous items. Modified after Griskevicius *et al.*[46]

quality to females, while females are more likely to signal their beneficence and proper maintenance of domestic life. Such signals are thought to have evolved because they indicate actual enhanced fitness and thus the quality of the potential mate. But Miller argues that selection in early hominid history was particularly driven by displays of the "central six" mental traits (intelligence, openness, conscientiousness, agreeableness, stability, extraversion), which are evolutionarily conserved, heritable, stable, measurable, reliable, and attractive to others.[47] Thus, Miller believes that consumerist capitalism evolved because of the human instinct to display intelligence and personality, which can be normatively expressed through certain positions and products, along with social and historical factors.

Converging both the life history strategy themes from Ellis' presentation and the evolutionary psychology themes of Miller's presentation, Vladas Griskevicius (assistant professor of marketing, University of Minnesota McKnight Land-Grant Professor) presented a life-history strategy approach to understanding short- versus

long-term strategies across socioeconomic groups. Griskevicius began with an evocative description of a man with a small income who works very hard in his job as a mechanic, yet spends up to $30,000 per year on lottery tickets. Such behavior appears risk seeking and irrational in the context of his other financial needs; but Griskevicius, like Ellis, used this case to portray the distinction between somatic and reproductive strategies for life investments. In this context, somatic strategies represent a longer horizon approach to improving physical health, longevity, and the accumulation of skills in situations where the environment appears secure, and reproductive strategies represent a shorter term strategy to compete for reproductive success in the present in situations where the environment appears risky and uncertain. Across many studies that utilize mortality primes from short newspaper articles about danger in society, Griskevicius *et al.* found that the fear of mortality causes bifurcating responses in individuals depending on their early (but not current) socioeconomic status (SES).[48] Individuals with early material and social support become risk averse when later primed to expect danger

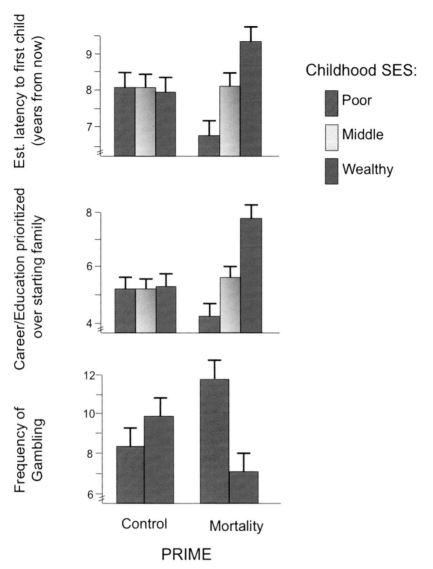

Figure 9. Effects of early childhood SES on the response to short- versus long-term planning strategy tasks. Participants from low childhood SES backgrounds (maroon) select the more immediate, short-term strategy, and those from high childhood SES backgrounds (blue) select the longer-term strategy (middle SES in gray bars). These effects generalize across decisions to predict the number of years before having one's first child (A), whether to prioritize starting a family or education and career (B), and whether to take an immediate smaller reward or to gamble for a longer-term, larger outcome (C). In graphs A and B, higher values on the *y*-axis represent greater slow, long-term investments, but in graph C, higher values represent greater immediate, fast-term investment. Modified after Griskevicius *et al.*[49]

or uncertainty, while those with early deprivation become risk seeking in the same later setting. These effects generalize across many measures, including the age at which people predict they will have their first child, the decision to start a family soon or to invest further in one's education or career, and the decision to take a smaller immediate reward versus a gamble for a larger, longer term benefit (Fig. 9). Even when individuals of low versus high childhood SES are exposed to the same cues of danger in the environment, those from a less enriched early environment switch into a faster, riskier strategy, while those from a more enriched environment switch to an even slower strategy. Thus, in a case such as that of the man who spent his small income on lottery tickets, a life-history strategy approach assumes that he is not irrational, but rather exhibits an evolved strategic response to the combination of his early and current environmental conditions.

Summary

Ecologists study optimal models of animal food hoarding, financial analysts model stock trading, psychologists investigate decisions about reward (and impairments therein such as gambling and addiction), and clinical psychologists treat compulsive hoarding. Each of these domains captures the ways in which decisions are made to acquire resources in order to balance short- and long-term needs. Progress in each domain can be catapulted by a genuine attempt to identify common themes and incorporate models from one domain into another. For example, by understanding the evolution of animal food hoarding, clinicians and marketers identify environmental triggers for human acquisition, yielding strategies that take natural tendencies into account. By bringing researchers of compulsive hoarding and shopping (which are sex biased, but no one knows why) together with neuroscientists, we can understand the neurobiology of these intractable disorders. Interactions between marketing, business, or finance with neuroscience can improve the sophistication of neural theories of choice, which are increasingly popular. Moreover, all of these fields can contribute to the dialogue on public policy.

In just a few short days, with scholars engaged in a dialogue across these domains, we were able to identify multiple commonalities across the findings of researchers, all of which addressed a general model of consumption as an evolved response to perceived cues of reward and uncertainty that differ across individuals due to inherited and learned sensitivities and strategies.

Maladaptive consumption is a serious issue that produces environmental waste, unfair labor practices, and negatively affects human health. Subunits of local and federal government separately struggle to encourage monetary saving, reduce waste, increase recycling, and deal with compulsive hoarding. Through a careful comparison of the mechanisms underlying these seemingly disparate processes, a unified model of resource allocation can be created that benefits basic science as well as society.

Acknowledgments

We are grateful to the committee, to Rackham Graduate School, and to the University of Michigan for their support in this endeavor. Peter Todd assisted both in decisions about the conference and in the preparation of this article.

Meeting financial contributors
Terri Lee and Therese Kummer (University of Michigan, Department of Psychology), and Randy Nesse (University of Michigan, Department of Psychology; Evolution and Human Adaptation).

Organizing committee members
R. Brent Stansfield (University of Michigan, Department of Medical Education), and Randy Nesse (University of Michigan, Department of Psychology), Susan Gelman (University of Michigan, Department of Psychology), Frank Yates (University of Michigan, Department of Psychology), Kent Berridge (University of Michigan, Department of Psychology), Colleen Seifert (University of Michigan, Department of Psychology), Peter Todd (University of Michigan, Department of Psychology), Richard Gonzalez and Fred Feinberg (University of Michigan, Department of Psychology), Corey Blant (University of Michigan, Department of Psychology), Brian Vickers (University of Michigan, Department of Psychology), Stephanie Carpenter (University of Michigan, Department of Psychology), David Chester (University of Michigan, Department of Psychology), and Rob Smith (University of Michigan, Department of Psychology).

Meeting support and assistance
Sue Schaefgen (University of Michigan Conference Services), Rick Richter (University of Michigan, Department of Psychology), Raye Holden (University of Michigan, CARSS Institute), Mary Mohrbach (University of Michigan, Department of Psychology), David Featherman (University of Michigan, CARSS Institute), Irv Salomeen (University of Michigan, CARSS Institute), and Carl Simon (University of Michigan, Center for Complex Systems, CARSS and Phoenix Institute).

Conflicts of interest

The author declares no conflicts of interest.

References

1. Schultz, W. 2002. Getting formal with dopamine and reward. *Neuron* **36:** 241–263.
2. Whishaw, I.Q. 1993. Activation, travel distance, and environmental change influence food carrying in rats with hippocampal, medial thalamic and septal lesions: implications for studies on hoarding and theories of hippocampal function. *Hippocampus* **3:** 373–385.
3. Whishaw, I.Q. & R.A. Kornelsen. 1993. Two types of motivation revealed by ibotenic acid nucleus accumbens lesions: dissociation of food carrying and hoarding and the role of primary and incentive motivation. *Behav. Brain Res.* **55:** 283–295.
4. Anderson, S.W., H. Damasio & A.R. Damasio. 2005. A neural basis for collecting behaviour in humans. *Brain* **128:** 201–212.
5. Hsu, M., M. Bhatt, R. Adolphs, *et al.* 2005. Neural systems responding to degrees of uncertainty in human decision-making. *Science* **310:** 1680–1683.
6. Knutson B. *et al.* 2007. Neural predictors of purchases. *Neuron* **53:** 147–156.
7. Saxena, S. *et al.* 2004. Cerebral glucose metabolism in obsessive-compulsive hoarding. *Am. J. Psychiatr.* **161:** 1038–1048.
8. Loewenstein, G. 1996. Out of control: visceral influences on behavior. *Organ. Behav. Hum. Decis. Process.* **65:** 272–292.
9. Richins, M.L. 1994. Valuing things: the public and private meanings of possessions. *J. Consum. Res.* **21:** 504.
10. Belk, R.W. 1988. Possessions and the extended self. *J. Cons. Res.* **15:** 139.
11. Richins, M.L. 1994. Special possessions and the expression of material values. *J. Consum. Res.* **21:** 522.
12. Cameron, E.H. 1923. The psychology of saving. *Ann. Am. Acad. Pol. Soc. Sci.* **110:** 156–164.
13. de Waal, F.B.M. 2009. *The Age of Empathy: Nature's Lessons for a Kinder Society.* Harmony Books. New York.
14. de Waal, F.B.M. & M. Suchak. 2010. Prosocial primates: selfish and unselfish motivations. *Phil. Trans. R. Soc. B* **365:** 2711–2722.
15. Naqvi, N.H. & A. Bechara. 2009. The hidden island of addiction: the insula. *Trends Neurosci.* **32:** 56–67.
16. Naqvi, N.H. *et al.* 2007. Damage to the insula disrupts addiction to cigarette smoking. *Science* **315:** 531–534.
17. Flagel, S.B., H. Akil & T.E. Robinson. 2009. Individual differences in the attribution of incentive salience to reward-related cues: implications for addiction. *Neuropharmacology* **56:** 139–148.
18. Yager, L.M. & T.E. Robinson. 2010. Cue-induced reinstatement of food seeking in rats that differ in their propensity to attribute incentive salience to food cues. *Behav. Brain Res.* **214:** 30–34.
19. Saunders, B.T. & T.E. Robinson. 2010. A cocaine cue acts as an incentive stimulus in some but not others: implications for addiction. *Biol. Psychiatr.* **67:** 730–736.
20. Flagel, S.B. *et al.* 2011. A selective role for dopamine in stimulus-reward learning. *Nature* **469:** 53–57.
21. Flagel, S.B. *et al.* 2007. Individual differences in the propensity to approach signals vs goals promote different adaptations in the dopamine system of rats. *Psychopharmacology* **191:** 599–607.
22. Knutson, B. & S.M. Greer. 2008. Anticipatory affect: neural correlates and consequences for choice. *Phil. Trans. R. Soc. B* **363:** 3771–3786.
23. Ellis, B.J. *et al.* 2009. The impact of harsh versus unpredictable environments on the evolution and development of life history strategies. *Hum. Nat.* **20:** 204–268.
24. Wilson, M. & M. Daly. 1985. Competitiveness, risk taking, and violence: the young male syndrome. *Ethol. Sociobiol.* **6:** 59–73.
25. Sherry, D.F. & J.S. Hoshooley. 2010. Seasonal hippocampal plasticity in food-storing birds. *Phil. Trans. R. Soc. B* **365:** 933–943.
26. Tolin, D.F. *et al.* 2011. Hoarding among patients seeking treatment for anxiety disorders. *J. Anxiety Disord.* **25:** 43–48.
27. Pertusa, A. *et al.* 2010. Refining the diagnostic boundaries of compulsive hoarding: a critical review. *Clin. Psychol. Rev.* **30:** 371–386.
28. Mataix-Cols, D. *et al.* 2010. Hoarding disorder: a new diagnosis for DSM-V? *Depress. Anxiety* **27:** 556–572.
29. Frost, R.O. 2010. Treatment of hoarding. *Expert Rev. Neurother.* **10:** 251–261.
30. Frost, R.O. & G. Steketee. 2010. *Stuff: Compulsive Hoarding and the Meaning of Things.* Houghton Mifflin Harcourt Publishing. New York.
31. Frank, R.H. 1999. *Luxury Fever: Why Money Fails to Satisfy in an Era of Excess.* The Free Press. New York.
32. Frank, R.H. 1988. *Passions within Reason: The Strategic Role of the Emotions.* W. W. Norton & Company. New York.
33. Lea, S.E.G. & P. Webley. 2006. Money as tool, money as drug: the biological psychology of a strong incentive. *Behav. Brain Sci.* **29:** 161–209.
34. Wang, J.M., R.D. Seidler, J.L. Hall & S.D. Preston. (under review). The neural bases of acquisitiveness: decisions to acquire and discard everyday goods differ across frames, items, and individuals.
35. Preston, S.D., J.R. Muroff & S.M. Wengrovitz. 2009. Investigating the mechanisms of hoarding from an experimental perspective. *Depress. Anxiety* **26:** 425–437.
36. Preston, S.D. 2001. Effects of stress on decision making in the Merriam's kangaroo rat (Dipodomys merriami). PhD Thesis, University of California, Berkeley.
37. Vohs, K.D. & T.F. Heatherton. 2000. Self-regulatory failure: a resource-depletion approach. *Psychol. Sci.* **11:** 249–254.
38. Vohs, K.D. & R.J. Faber. 2007. Spent resources: self-regulatory resource availability affects impulse buying. *J. Consum. Res.* **33:** 537–547.
39. Vohs, K.D., R.F. Baumeister & N.J. Ciarocco. 2005. Self-regulation and self-presentation: regulatory resource depletion impairs impression management and effortful self-presentation depletes regulatory resources. *J. Pers. Soc. Psychol.* **88:** 632–657.
40. Vohs, K.D. 2006. Self-regulatory resources power the reflective system: evidence from five domains. *J. Consum. Psychol.* **16:** 215–221.
41. Vohs, K.D. *et al.* 2008. Making choices impairs subsequent self-control: a limited-resource account of decision making,

self-regulation, and active initiative. *J. Pers. Soc. Psychol.* **94:** 883–898.

42. Vohs, K.D. & B.J. Schmeichel. 2003. Self-regulation and the extended now: controlling the self alters the subjective experience of time. *J. Pers. Soc. Psychol.* **85:** 217–230.

43. Schmeichel, B.J., K.D. Vohs & R.F. Baumeister. 2003. Intellectual performance and ego depletion: role of the self in logical reasoning and other information processing. *J. Pers. Soc. Psychol.* **85:** 33–46.

44. Schmeichel, B.J. & K.D. Vohs. 2009. Self-affirmation and self-control: affirming core values counteracts ego depletion. *J. Pers. Soc. Psychol.* **96:** 770–782.

45. Miller, G. 2009. *Spent: Sex, Evolution, and Consumer Behavior*. Viking. New York.

46. Griskevicius, V. *et al.* 2007. Blatant benevolence and conspicuous consumption: when romantic motives elicit strategic costly signals. *J. Pers. Soc. Psychol.* **93:** 85–102.

47. Penke, L., J.J.A. Denissen & G.F. Miller. 2007. The evolutionary genetics of personality. *Eur. J. Pers.* **21:** 549–587.

48. Griskevicius, V. *et al.* 2011. The influence of mortality and socioeconomic status on risk and delayed rewards: a life history theory approach. *J. Pers. Soc. Psychol.* **100:** 1015–1026.

49. Griskevicius, V. *et al.* 2011. The influence of mortality and socioeconomic status on risk and delayed rewards: a life history theory approach. *J. Pers. Soc. Psychol.* **100:** 1015–1026.

Ann. N.Y. Acad. Sci. ISSN 0077-8923

ANNALS OF THE NEW YORK ACADEMY OF SCIENCES
Issue: *Annals Meeting Reports*

Wild immunology: converging on the real world

Simon A. Babayan,[1] Judith E. Allen,[1] Jan E. Bradley,[2] Markus B. Geuking,[3] Andrea L. Graham,[4] Richard K. Grencis,[5] Jim Kaufman,[6] Kathy D. McCoy,[3] Steve Paterson,[7] Kenneth G. C. Smith,[8] Peter J. Turnbaugh,[9] Mark E. Viney,[10] Rick M. Maizels,[1] and Amy B. Pedersen[1]

[1]Centre for Immunity, Infection and Evolution, Institutes of Immunology and Infection Research, and Evolutionary Biology, School of Biological Sciences, University of Edinburgh, Ashworth Laboratories, Kings Buildings, Edinburgh, United Kingdom. [2]School of Biology, University of Nottingham, Nottingham, United Kingdom. [3]Department of Clinical Research, Universitätsklinik für Viszerale Chirurgie und Medizin Inselspital, University of Bern, Bern, Switzerland. [4]Department of Ecology and Evolutionary Biology, Princeton University, Princeton, New Jersey. [5]Faculty of Life Sciences, University of Manchester, Manchester, United Kingdom. [6]Department of Pathology (and Department of Veterinary Medicine), University of Cambridge, Cambridge, United Kingdom. [7]Institute of Integrative Biology, University of Liverpool, Liverpool, United Kingdom. [8]Cambridge Institute for Medical Research and the Department of Medicine, University of Cambridge School of Clinical Medicine, Addenbrooke's Hospital, Cambridge, United Kingdom. [9]FAS Center for Systems Biology, Harvard University, Cambridge, Massachusetts. [10]School of Biological Sciences, University of Bristol, Bristol, United Kingdom

Address for correspondence: Amy B. Pedersen, Advanced Fellow, Centre for Immunity, Infection and Evolution, Institutes of Evolutionary Biology, Immunology, and Infection Research, School of Biological Sciences, University of Edinburgh, Kings Buildings, Ashworth Labs, West Mains Road, Edinburgh EH9 3JT, UK. amy.pedersen@ed.ac.uk

Recently, the Centre for Immunity, Infection and Evolution sponsored a one-day symposium entitled "Wild Immunology." The CIIE is a new Wellcome Trust–funded initiative with the remit to connect evolutionary biology and ecology with research in immunology and infectious diseases in order to gain an interdisciplinary perspective on challenges to global health. The central question of the symposium was, "Why should we try to understand infection and immunity in wild systems?" Specifically, how does the immune response operate in the wild and how do multiple coinfections and commensalism affect immune responses and host health in these wild systems? The symposium brought together a broad program of speakers, ranging from laboratory immunologists to infectious disease ecologists, working on wild birds, unmanaged animals, wild and laboratory rodents, and on questions ranging from the dynamics of coinfection to how commensal bacteria affect the development of the immune system. The meeting on wild immunology, organized by Amy Pedersen, Simon Babayan, and Rick Maizels, was held at the University of Edinburgh on 30 June 2011.

Introducing wild immunology

In the wild, organisms face many pressures (e.g. parasites, pathogens, commensal organisms, seasonality, resource availability) that affect their health and fitness. A great deal of our knowledge on infection and immunity, however, has been developed in highly controlled laboratory settings where variation is minimized to more easily identify molecular and cellular immune mechanisms. While traditionally maintaining strong connections between the lab and the field, greater practical achievements in human medicine and veterinary medicine have been difficult because of the challenges of associating detailed mechanistic interventions developed in the lab with what has been learned in the field about the effects on host fitness in natural settings. The aims of the conference on wild immunology were to bring these approaches together, to address shared questions about the role of the immune system in natural populations, and to better integrate laboratory-based immunology into the actual health of humans and animals.[1]

Judi Allen (professor of immunobiology, University of Edinburgh) introduced the topic of wild immunology from her perspective as a laboratory-based immunologist. She argued that wild immunology draws from both ecological immunology (eco-immunology) and laboratory-based immunology, and aims to link immunity and infection with host health and fitness in wild systems.[1] Specifically, Allen said, wild immunology is "about

doi: 10.1111/j.1749-6632.2011.06251.x

communicating between lab studies and studies in *wild* systems."

The burgeoning field of eco-immunology has focused on understanding costs and trade-offs in an organism's life-history and fitness associated with mounting immune responses (see, for example, Refs. 2 and 3). Many studies in this field have aimed to identify a single measure of the immune response (e.g., white blood cell number[4] or spleen size[5]), termed *immunocompetence*, as a meaningful metric of immune function and the relationship of immune function to host fitness or disease burden. This approach may be counterintuitive to those laboratory immunologists who instead seek to understand the complex network of all, or the majority, of the mechanisms that underlie immune functions in general. The link and yet greatest difference between immunologists and ecologists is in their approach to variation (e.g., host genotype, parasite genotype, sex, seasonality, coinfection, and commensal community); laboratory immunologists aim to minimize variation ("noise") in the systems they study, while ecologists focus squarely on natural variation. Generally, ecologists have created a toolbox with which to understand and explain variation in wild systems, and this is often at the heart of their research. In contrast, laboratory immunologists intentionally reduce variation by controlling, for example, the genetics, environment, and resources in experimental systems in order to pinpoint important underlying mechanisms. Allen suggested that it is time for both fields to "get real," that is, ecologists need more specific measures of immune competence that are based on *real* mechanisms of immunity and resistance, while laboratory immunologists need to get out of the lab to understand if the mechanisms they define in controlled lab systems are *real*—meaningful—for health and fitness of wild systems.

Collaborations across this divide can provide unique and valuable insight (for example, see Refs. 1 and 6). Currently, however, several barriers exist that prevent successful interdisciplinary communication, such as the different languages and approaches employed by ecologists and laboratory immunologists. Allen suggested, among other things, that an important divide separating ecologists and laboratory immunologists is the ways they write and read scientific literature. For example, most ecological/evolutionary studies provide essential informa-

tion on the study design, with extensive reference to the statistical/modeling approaches undertaken; such information is a central element of Materials and Methods sections of papers in this field, with ecologists scouring these sections. By contrast, laboratory immunologists often read Materials and Methods sections of manuscripts only if they want to know the details of a particular experiment, and typically they do not consider which statistical methods were employed as central to the paper. Additionally, there are basic language and communication differences that make successful crosstalk difficult. For example, ecologists define *tolerance* as a host's ability to limit damage caused by a given parasite density.[7–9] In contrast, laboratory immunologists define tolerance as the process (or processes) by which immune responses (to an antigen, including self-antigens) are negatively regulated;[10] thus, a weaker immune response can be associated with lower risk of immunopathology. In such counterintuitive cases, especially, proactive discussions between ecologists and laboratory immunologists are likely to prove the most compelling and productive.

From the field to the lab

The next talk by Jan Bradley (professor of parasitology, University of Nottingham) focused on using wild rodents as model systems for understanding mammalian immune systems in a naturally variable world. Humans and wild animals live in variable environments and are exposed to, and often infected by, a diverse array of commensals, parasites, and pathogens throughout their life. Bradley pointed out that, in humans, coinfections can modulate immune responses and affect regulatory processes.[11] Yet, Bradley suggested that understanding how variation affects immune responses in humans is difficult because of ethical and financial issues. However, she highlighted that by adapting the tools and techniques from laboratory-based immunological studies on inbred mice to wild rodents may be a useful avenue for understanding how variation affects immunity.[12] Specifically, measuring immune phenotypes, which will reflect immune responsiveness and function, of wild rodents within the context of their environments, genotypes, and pathogens may provide an excellent opportunity to uncover how the mammalian immune system responds to diverse infections in the natural environment.

In the first part of her presentation, Bradley discussed recent research on innate immune responses in wild rodents, specifically of the wood mouse *Apodemus sylvaticus*, a native rodent of UK woodlands. Using a cross-sectional study design, Bradley and colleagues collected wild wood mice, assessed their macro- and microparasite infections, and conducted a series of innate immunological analyses. Specifically, they used a panel of Toll-like receptor (TLR) agonists to stimulate immune responses in splenocytes, the surrogate marker for which was production of tumor necrosis factor alpha (TNF-α). They found that burdens of the intestinal nematode *Heligomosomoides polygrus* and the louse *Polyplax serrata* were negatively correlated with TNF-α production.[13] By contrast, the response of intestinal protozoan microparasites *Eimeria spp.* to TLR agonist stimulation was significantly positively associated with TNF-α production, the most important TLR receptors being TLR7 and TLR9.[13] Such TLR-mediated responses, likely meaningful indicators of innate responsiveness in general,[14] may be an important tool for understanding the immunomodulatory effects of parasites in wild systems. Bradley and colleagues are now building on these results from wild mice by working back in the lab using multiple low-dose ("trickle") nematode infections to better model real-world exposure in lab mice.

The next part of the talk focused on investigating the immune responses of an extensively studied population of wild field voles (*Microtus agrestis*) in Kielder Forest in Northern England. The vole populations fluctuate on a 3- to 4-year cycle,[15] and much is known about their microparasite (e.g., Cowpox virus,[16] *Trypanosoma microti*,[17] *Bartonella spp.*,[18] and *Anaplasma phagocytophilum*[19]) and macroparasite[20] communities. Again using a cross-sectional study design, Bradley and colleagues measured several cellular assays of pro- and anti-inflammatory signaling responses (both transcription factors and cytokines) quantified by Q-PCR. They found striking variation in immune measures among the field voles, specifically that temporal patterns (i.e., seasonality) and life history stages (i.e., reproduction) strongly affected immune expression.[21] Strong negative associations between inflammatory mediators and measures of condition (i.e., liver size, spleen size, and splenocyte number) were also found. Using a grouping analysis, Bradley and colleagues found that measures of pro- and anti-inflammatory responses were highest in the winter but differed in summer when the minimum responses were found.[21]

Bradley highlighted that these novel approaches will help build useful immunological tools and techniques to begin to understand the variation in immune phenotypes found in wild populations. By integrating the various methods (measuring life history, parasite infections, environmental variation, and immune phenotypes) within both observational and, eventually, experimental approaches, we will begin to understand what factors drive immune phenotypes and parasite burdens in wild populations. Bradley concluded by suggesting that wild immunology will be a very useful avenue that links the laboratory mouse with the real world variation seen in human populations.

Immune potency of wild mice

Mark Viney (professor of zoology, University of Bristol) presented a comparative study of immune function in wild and laboratory mice. Viney suggested that the immune responses of wild animals have been rather poorly studied despite the fact that lab and wild rodents often vary in their response to parasites and pathogens.[1] He highlighted that the immune function of wild animals may differ from that of laboratory-bred animals because of their different environments. Viney contends that this idea follows from the concept in life history theory called *resource partitioning*, in which animals distribute scarce resources to all aspects of life, including costly immune responses. Therefore, well-resourced laboratory animals in benign, or controlled, environments may have substantially greater resources to direct to immune function than wild mice, which live in demanding environments where resources can be much more limited.[22] A logical extension of this idea is that there may be substantial inter-individual variation in the immune function of wild animals because different individuals acquire different amounts of resources and have different demands on expenditure of those resources.

To investigate this idea, Viney and colleagues compared the immune function of a laboratory bred mouse strain (C57BL/6, a widely used strain that has been maintained for many decades) and wild-caught *Mus musculus*.[23] Specifically, they compared the immune responses of lab and wild-caught mice to a novel antigen with which they were immunized and found that, by most immune measures, wild-caught mice produced higher concentrations and more avid antigen-specific IgG responses, as well as higher concentrations of total IgG and IgE, compared with laboratory-bred mice. Viney also analyzed cell populations by flow cytometry, which demonstrated a comparatively greater overall level of activation of T helper cells, macrophages, and dendritic cells in wild mice, but no differences in regulatory T (T_{reg}) cell activation.[23] Importantly, they observed that these immune measures were substantially more variable among wild-caught mice than among the laboratory-bred mice (as in Ref. 24). Thus in contrast to their original hypothesis about limited resources in the wild leading to poorer immune function, these results suggest that wild mice have stronger and more varied immune responses, despite being less than two-thirds the mass of lab mice and having higher levels of leptin (a hormone involved in energy expenditure and metabolism).

The next challenge of this research will be to understand which aspects of an individual animal's life determines its immune function in the wild and what factors lead to the extreme variation found across individuals. Lastly, Viney pointed out that the study of wild immunology is important because it allows us to link measures of immune function with host health and fitness in natural settings and expands the current taxonomically restricted set of laboratory animals used in immunological studies.

Cytokine polymorphisms in wild populations

Steve Paterson (senior lecturer, University of Liverpool) presented a study linking polymorphisms in cytokine-coding genes with immune responses and pathogen resistance in wild rodent populations. His research was motivated by a major question in wild immunology: why do individuals vary in their immune responses and resistance to pathogens and parasites? Many of the sources of this variation may be environmental, such as nutrition, microbiota (microorganisms inhabiting the gastrointestinal tract), and coinfection.[1,23] Equally, however, an important source of variation in immune responses and resistance is likely to be due to genetic differences among individuals. In support of this, some individuals in human populations appear to be far more susceptible to helminth reinfection than

others, which may be correlated to distinctive immunological phenotypes.[25–27] Quantitative genetic analysis of parasite infection and reinfection also highlights a strong genetic component to susceptibility.[28,29]

However, until now, genetic studies of immune response and resistance in natural populations have been limited almost entirely to the major histocompatibility complex (MHC).[30,31] This is an important, highly polymorphic component of the vertebrate immune system required for antigen presentation. Paterson highlights that despite numerous studies having demonstrated balancing selection from sequence and allele frequency data, and linked individual variation to differences in pathogen resistance, they have been almost exclusively based on the MHC. However, the MHC only contributes a small proportion of the genetic variation in resistance to infectious disease in mammals.[32] Indeed, the MHC is only a subset of the genes involved in the immune system, and Paterson and colleagues argue that molecular ecologists have been somewhat narrow in not looking for functional variation elsewhere in the genome with respect to pathogen resistance. Because of this, molecular ecology lags behind human genetics, where a far broader range of candidate loci are considered and where genome-wide association studies (GWAS) are now increasingly being deployed to understand the genetic determinants of pathogen resistance and immunity.[33]

Cytokines are key signaling molecules of the immune system, and genetic polymorphisms in cytokine-coding genes have repeatedly been implicated in resistance to pathogens in humans.[34–36] As such, they potentially provide a group of genes that may underlie variation in resistance to a wide range of infections in the natural environment. Paterson used the same wild vole system as described above by Bradley, where individuals live in natural conditions and are likely exposed to a diverse array of parasites. His group conducted a single nucleotide polymorphism (SNP) analysis of the cDNA produced from a wide spectrum of cytokine genes to identify signatures of selection acting on these polymorphisms, and to associate cytokine polymorphisms with variation in immunological parameters and pathogen resistance. They found that two cytokine genes, *Il1b* and *Il2*, showed evidence of balancing selection acting on both DNA sequence and allele frequencies.

Using sequence-based tests, they also showed low genetic differentiation between various field sites but strong associations between cytokine polymorphisms and levels of infection. *Il1b, Il2,* and *Il12* were associated with variation in a number of immunological parameters (e.g., IL-1b, IL-10, and IL-2 expression) and, in turn, with variation in resistance to multiple natural pathogens.

Paterson's results demonstrate that cytokines represent an important source of genetic variation for resistance to a range of pathogens in wild systems and that this genetic component was as important as sex and age in predicting infection status. He highlights the potential for future work to extend the range of genes studied—in tandem with further immunological assays—in order to understand how genetic diversity leads to functional variation in pathogen resistance in natural populations and, in turn, how such genetic diversity is maintained. In this respect, Paterson concluded by suggesting that wild immunology of rodents may be an informative model for the natural variation in immunity found in human populations.

Autoimmunity and antibodies in the balance

Andrea Graham (assistant professor of ecology and evolutionary biology, Princeton University) spoke about using antibodies to understand immunoheterogeneity among the Soay sheep of St. Kilda in the Outer Hebrides. As suggested by Graham, antibodies are important immune effector molecules for a variety of infectious diseases, and they also present logistical advantages for studies of immunology in the wild. These advantages include relative stability *in vivo* and in the freezer; accurate measurement of antibodies requires small sample volumes (usually <5μl), which are more logistically feasible for many wild systems; and different antibody isotypes permit inference about the cytokine milieu in which a response is induced (e.g., Refs. 37 and 38), and, in principle, a wide array of antibody specificities—for example, to self versus nonself antigens, or across a range of parasite species and strains—are accessible for analysis. However, not all infections are cleared by antibodies and off-the-shelf reagents are not available for all host species, especially wild animals. Graham pointed out that investigators must therefore decide whether antibody measurements or other biomarkers are most suitable for the question and host–parasite system under study.[9]

The Soay sheep of St. Kilda present a wild system in which measurement of antibodies is both feasible and potentially informative. For example, thanks to the extensive demographic and genetic data collected during longitudinal study of the sheep,[39] antibody measurements can be analyzed in a whole-organism context, including assessment of associations between antibody titers and fitness.[6] The sheep are infected by and, in harsh winters with low food availability, are killed by intestinal nematodes.[40] Annual plasma samples are taken and veterinary immunological reagents that work in sheep are available.

The initial incentive for Andrea Graham, Dan Nussey, and colleagues to measure antibodies of the Soay sheep was to ask whether animals with limited food and abundant (possibly immunosuppressive) helminth infections exhibit autoimmunity. Thus, the first plasma molecules they chose to measure were antinuclear antibodies (ANA) that bind nuclear and cytoplasmic constituents of mammalian cells (for example, histones or tRNA). ANA are considered risk factors for systemic autoimmune diseases such as lupus.[41] However, ANA also include infection-protective natural antibodies[42] and/or may indicate highly responsive B cells.[43] Indeed, clinical evidence suggests that only after many years at high titers are ANA associated with disease.[41] It is currently unknown how wild animals balance effective parasite killing immune effector mechanisms, while limiting damage to self.

Graham's results suggest that despite food limitation and high parasite burdens, some Soay sheep were ANA+ (based on clinical criteria). More interestingly, ewes with high ANA concentrations lived significantly longer than ewes with low ANA but reduced annual reproduction.[6] Due to these balancing associations, and possible trade-offs between survival and fecundity, the data suggest how immunoheterogeneity might be maintained in the wild and confirm a major prediction of life history theory.[44] However, any links between autoimmunity and resistance against infection (such as described by Ken Smith during the symposium; e.g., Ref. 45) in the Soay sheep are not yet clear. It is critical to forge such links, so Graham and colleagues are currently investigating relationships among titers of ANA, parasite-specific antibodies, and natural antibodies—as well as the potential for maternal antibody transfer—to explain variation in survival among these sheep. Antibodies are protective against several developmental stages of the nematode *Teladorsagia circumcincta*,[46] which is among the most pathogenic parasites infecting the Soay sheep.[40]

The presentation highlighted that wild "model" systems, such as the Soay sheep of St. Kilda, provide fascinating new insights that are not provided by laboratory model systems. Graham concluded by suggesting that although the genetic and environmental variability of wild systems render detailed understanding of immune mechanisms difficult to obtain, they do permit analysis of how immune systems function, and why they vary, in the face of multiple natural challenges.

Immunological evolution

Jim Kaufman (professor of comparative immunogenetics, University of Cambridge) discussed how studies of the immune response in birds and other non-mammalian vertebrates can tell us about pathogens, coevolution, and genomic organization. Kaufman highlighted that immune responses to most infectious or immunization challenges are enormously complex, presumably due to many millions of years of evolution that involved step-by-step acquisition of protective mechanisms adapted from different molecular and cellular processes, with each step responding to specific challenges. Kaufman suggested that while it is interesting to understand fundamental processes and principles underlying the evolution of immune responses, reconstructing the events that have occurred has been a difficult task. However, one component that is particularly amenable to such analysis is the adaptive immune system of vertebrates, which is clearly a *system* with a unique origin.[47] He referenced research on mammals by many groups over the last 40 years that have revealed key immune genes (such as genes encoding MHC, TcR, and antibodies), as well as the associated cells and processes that had already evolved at the emergence of the jawed vertebrates (see Ref. 47).

While many insights about the immune response have come from mammals, chickens have been a key organism for understanding the origin and evolution of the adaptive immune system.[48] Kaufman pointed out that advantages of the chicken, compared to non-mammalian vertebrates, are primar-

ily due to the economic and social importance of poultry as a food source. There is an enormous global poultry industry driving intense scrutiny of pathogens, genetics, and genomics, and immune responses to both diseases and vaccines. In addition, a wide range and huge number of chickens live in seminatural conditions and allow easy movement from field to lab and back again. Kaufman and colleagues have tried to understand why, in contrast to mammals, the chicken MHC has such strong genetic associations with resistance to infectious pathogens and responses to vaccines. They have found that chickens have single dominantly expressed class I and class II molecules, the properties of which can determine the immune response.[49,50] The basis for the dominantly expressed class I gene was due to coevolution with highly polymorphic TAP and tapasin genes located nearby.[51,52] In addition, surprisingly, NK receptor (NKR) genes are also present in the chicken MHC genomic region.[52–54]

The salient features that Kaufman and colleagues discovered are found in many, if not most, non-mammalian vertebrates, which suggests that the ancestral organization was like that of chickens, with the mammalian MHC arising by a "messy" inversion.[47,48,55] The presence of NKR genes suggests that ancestors of the receptors (such as NKR and T cell receptors) were also present in the primordial MHC, most likely in order to co-evolve with their ligands into a functioning system. Many other disparate data are explained by, and support, the view that the primordial MHC is the birthplace of the adaptive immune system.[47,48,56]

Chickens offer certain advantages to understanding the function and origin of the immune system over inbred laboratory mouse strains and outbred people. However, there are several issues for which wild animals, living without direct human intervention, would be more suitable. In particular, Kaufman suggests that the domestic chicken, even in the roughest barnyard setting, would not be an adequate system in which to understand immune responses of wild animals within the context of natural and complex ecosystems. Moreover, understanding the extent to which the salient features discovered in the chicken are general to other non-mammalian vertebrates can only be determined by studying many other species, most of which would be wild. Kaufman said that there is every reason to expect many exciting surprises in this quest but cautioned that there are at least three groups working on the features of the MHC (biomedical scientists, farm animal health researchers, and evolutionary biologists), whose goals, methods, literatures, and citations are almost non-overlapping. In conclusion, Kaufman highlighted that increased dialogue between wild and lab studies can only improve matters, and provide novel insights into host–pathogen coevolution and the variations possible in the evolution of of the immune response.

The human dimension: polymorphism and autoimmunity

Ken Smith (professor of medicine, University of Cambridge) spoke about genetic predisposition to systemic lupus erythematosus (SLE), which is four to eight times more prevalent in people of African and Asian descent compared to people from European descent, and he presented evidence that this may, at least in part, result from selection for resistance to infection, in particular resistance to malaria. The risk of SLE is regulated by a large number of susceptibility alleles. A number of GWAS in thousands of lupus patients have identified over 35 loci that contribute to disease, but nonetheless explain only a relatively small proportion of the genetic contribution to SLE susceptibility.[57] One particularly important locus for SLE, implicated in both mouse and human, is on chromosome 1 and contains the low-affinity FcR (receptors that bind the Fc region of immunoglobulins) genes. The complexity of this region, due to it having arisen by duplication events and being prone to common copy number variation,[58] has resulted in it being largely excluded from SNP-based analyses.

Fcγ receptors are important regulators of inflammatory responses to antigen–antibody immune complexes. They include both high- and low-affinity receptors, all of which have an activation function except for the inhibitory FcγRIIB (CD32B), which is expressed by B cells, dendritic cells (DCs), macrophages, activated neutrophils, mast cells, and basophils.[59] Importantly, FcγRIIB is the only Fcγ receptor expressed on B cells in both humans and mice, and it controls many aspects of the antibody response, providing a form of feedback inhibition.

FcγRIIB is found in an important but complex susceptibility locus for SLE in both mice and humans. Mice deficient in the receptor are prone to autoimmunity.[59,60] Consistent with this, mice over-expressing FcγRIIB are protected from autoimmunity;[61] FcγRIIB also plays an important role in defense against bacterial infection.[61,62] Naturally occurring polymorphisms in the promoter region of FcγRIIB have been found to be associated with reduced receptor expression and with autoimmune-prone strains,[63,64] but the variant most associated with SLE is very common in wild mice, suggesting an evolutionary advantage. This wild mouse variant has been "knocked-in" to a conventional mouse strain, indicating how subtle changes in FcγRIIB regulation can drive autoimmunity (Espeli and Smith, in preparation). Human polymorphisms in FcγRIIB include mutations in promoter regions that may control expression levels, but these have not been studied in detail.[59]

A further polymorphic allele in humans is a coding variant in the transmembrane domain (I232T). This allele, first associated with SLE in Japanese populations[65] (and subsequently shown to hold true in Europeans[58]), abolishes the inhibitory activity of FcγRIIB.[66,67] Smith and others showed that this allele is homozygous in only around 1% of Europeans and between 5–10% of Africans and East Asians, ethnic groups known to suffer a higher incidence of SLE (at least when residing in the developed world). This association implies that an infectious agent might provide the balancing selection that would maintain the SLE-associated mutation at higher frequencies within some human populations.

This possibility was first tested in mice, and FcγRIIB deficiency was found to confer resistance to *Plasmodium chabaudi* malaria through increased phagocytosis, inflammatory cytokine production, and antibody titers.[45] These results were later confirmed in experiments with lethal strains of rodent malaria.[68] Primary human phagocytes isolated from volunteers homozygous for the inactive I232T variant of FcγRIIB showed heightened phagocytosis of *Plasmodium falciparum*–infected red blood cells,[45] which supports the hypothesis that a trade-off between resistance to malaria and propensity to develop SLE may be mediated by the FcγRIIB locus.[45] To further test their hypothesis in the field, two cohorts of children with severe malaria from Kilifi in Kenya were genotyped, and homozygosity for T232 was associated with substantial protection against severe malaria, with an odds ratio of 0.56, similar to that conferred by heterozygous thalassaemia.[69]

These studies from Smith *et al.* are remarkable in that epidemiological observations made in humans were traced back, owing to their homologies in laboratory mice, to specific alleles that may explain the maintenance of autoimmune disease in humans as a consequence of selective pressures imposed by pathogens.

Whipworm from the wild

Richard Grencis (professor of immunology, University of Manchester) spoke about the importance of the lab-generated mechanistic understanding of immunity to the whipworm, *Trichuris muris*. Specifically he highlighted the role of infective dose on the establishment of chronic infections and of the difficulty of deriving meaningful immunological readouts associated with patterns of infection in multiple infections. Whipworms are found in virtually every terrestrial mammal and are common in most wild rodents, reaching, for example, 88% prevalence in house mice. In 1954, *Trichuris muris* eggs were isolated from the feces of *Mus musculus* obtained from the birdhouse in the Zoological Park in Edinburgh.[70] Following embryonation, successful infection of laboratory mice was achieved, with low numbers of mature worms present in the caecum. Grencis pointed out that this strain of *T. muris* has subsequently been maintained successfully in laboratory mice in various laboratories around the world and has been used extensively as a system to study the host–parasite relationship with an emphasis on the immune response. The lab strain has helped generate the paradigm of Th2-mediated resistance to gastrointestinal helminths and dysregulation of this response leading to chronic infection.[71] In addition, Grencis highlighted that studies of this parasite strain have played a key role in identifying novel mechanisms of host protection and immunoregulation. Moreover, *T. muris* serves very well as a model of human Trichuriasis.

A small number of studies have explored *T. muris* infection in the wild and examined controlled infection in wild mice or outbred laboratory mice. In summary, data from field studies show that infection is generally characterized by low burdens of adult parasites (<20 worms), with only a few mice harboring many parasites. Grencis showed evidence that both laboratory-infected wild mice and outbred mice expelled their worm burden efficiently, although low-dose infections progressed readily to

patency. Further studies followed trickle infections in outbred mice in which repeated low-dose infections were given to mimic the manner in which animals in the field are likely to acquire infection. These studies demonstrated that low doses of eggs led to patent low-level infections, but for most mice, parasite numbers did not keep increasing, suggesting that some form of resistance against incoming parasites was operating.[72,73] Grencis suggested that the immunological basis of these observations only became apparent with the definition of resistance and susceptibility using inbred mice. Such studies defined key times post-infection that correlated with particular immune markers (such as cytokines, antibodies, and immunopathological features) and resistance or susceptibility to chronic infection. A series of experiments using low- or high-dose infections, given singularly or in multiples, including trickle infections, were subsequently carried out in inbred mice.[74] Grencis argued that the data clearly showed that when a single infection is given, assessment of immunological parameters is a very useful predictor of resistance status in terms of parasite burden. However, following complex infection regimes, such as trickle infection, the accuracy of these parameters as predictors of worm burden is markedly reduced, even in studies in which all other variables, such as sampling time, coinfection, and nutritional status, were carefully controlled. Nevertheless, in terms of worm burdens, there was evidence that some resistance could be built up over time in a manner similar to studies in wild mice.

Overall, the use of laboratory experiments with *T. muris* has been instrumental in defining mechanisms of resistance and susceptibility to intestinal nematode infection. Moreover, manipulation of infection burdens and regimes in inbred mice can reflect, at least in part, the complex situation that is experienced in the wild. Grencis finally drew attention to the fact that such studies do show, however, that inferring resistance status from analysis of immune parameters (known to play functional roles in well-defined conditions) remains a considerable challenge for the wild immunologist.

Microbes in the picture

Markus Geuking (research fellow, University of Bern) in collaboration with Kathy McCoy (professor, University of Bern) presented their work on the maintenance of intestinal homeostasis in response to controlled colonization with commensal bacteria. Geuking began his presentation by reminding the audience that humans are born germ free and then subsequently colonized with commensal bacteria. The highest density of bacteria ($>10^{11}$ CFU/gram) is found in the colon, and these bacteria are kept from entering the host by a single layer of epithelial cells covered by mucus—as the only physical barrier. Interestingly, humans have over 50% of all immune cells located in the intestine. Geuking asks how the relatively peaceful coexistence of bacteria and immune cells is maintained.

Vertebrates have coevolved with the community of intestinal microbes (microbiota), and they usually coexist in peaceful mutualism. Although these bacteria carry powerful inflammatory patterns that can be recognized by Toll-like receptors and other pattern recognition receptors, the intestinal immune system has adapted to the presence of these bacteria and does not, in healthy individuals, induce an inflammatory response. To study the mechanisms of intestinal immune adaption to the presence of commensal bacteria, Geuking and colleagues use axenic (germ-free) and gnotobiotic (defined microbiota) mouse models.[75] This allows them to control and define the microbiota present, as well as to manipulate the host immune system by using genetically modified mouse strains that are re-derived to germ-free status by axenic two-cell embryo transfer before being colonizing with a defined microbiota.

Several studies have investigated the immune responses to individual bacterial species.[76–79] To study the adaptation of the intestinal CD4$^+$ T cell compartment to colonization with a truly mutualistic commensal microbiota, Geuking *et al.* colonized germ-free mice with the defined altered Schaedler flora, which consists of eight species. They found that commensal colonization induced a strictly compartmentalized T$_{reg}$ cell response in the colon lamina propria.[80] This T$_{reg}$ response was functional because the same benign microbiota induced Th1 and Th17 effector responses in a mouse strain with defective T$_{reg}$ cells (due to transgenic expression of a virus-specific T cell receptor), which were controlled by transferred wild-type T$_{reg}$ cells.

Geuking concluded his talk by underlining the importance of immunological adaptations to the microbiota in maintaining gut homeostasis. He suggested that given that colonization of "true" commensals induces a regulatory T cell response that controls proper adaptation of the intestinal CD4$^+$ T cell compartment, it is likely that more complex microbiota that also harbor potential opportunistic bacteria and parasites, as seen in wild hosts, may determine the ability of hosts to generate immune responses against pathogens.

A wild world within

Peter Turnbaugh (Bauer Fellow, Harvard University) presented recent research exploring the differences and similarities of the gut microbiomes of laboratory and wild mammals. Humans and other mammals have coevolved with trillions of microorganisms, whose aggregate genomes often contribute functions not encoded by our own human genes.[81] The largest collection of these microorganisms, whose genomes are together referred to as the gut microbiome, is found throughout the gastrointestinal tract and can have a profound influence on health and disease, such as affecting nutrition and energy balance (i.e., diabetes, obesity, and metabolic syndrome, e.g., Refs. 82 and 83), inflammatory bowel disease,[84,85] immune development,[78] cardiovascular disease,[86] and xenobiotic metabolism.[87,88] However, unlike the human genome, the gut microbiome can be shaped by a variety of environmental factors, such as drug use,[89] diet,[90] probiotics,[91] and delivery method.[92] Turnbaugh highlighted that this plasticity, coupled with physiological relevance, makes the gut microbiome an attractive target for personalized medicine and/or nutrition.

To date, many of the studies of the mammalian gut microbiome have focused on inbred mouse lines in controlled laboratory settings. For example, comparisons of obese and lean mice have demonstrated differences in their gut microbiomes, including an increased relative abundance of genes for dietary energy harvest.[93,94] Interestingly, these "obesity-associated" communities can be used to transmit host adiposity: recipient germ-free mice colonized with a sample taken from an obese donor have a significantly greater gain in adiposity than mice colonized from a lean donor.[93,94] In an attempt to bridge the gap between laboratory mice and humans, Turnbaugh and colleagues have recently turned to "'humanized mice": formerly germ-free animals colonized with a human donor sample.[90] These animals are then used to investigate the distribution of bacteria throughout the length of their gastrointestinal tract and the succession of the gut microbial community early in life. Comparisons of humanized mice fed a low-fat polysaccharide-rich chow diet or a high-fat/high-sugar "Western" diet emphasize that dietary shifts can have a consistent, rapid, and significant impact on microbial community structure, gene content, and gene expression.[90]

Turnbaugh highlighted the fact that humans, the most commonly studied "wild animal," show similar shifts in microbial community structure and gene content when comparing obese and lean individuals,[95,96] although recent studies suggest that this relationship may be complicated by a variety of host and environmental factors.[97] Of note, a large-scale analysis recently identified three functional modules—including ATPases— in the gut microbiome that were correlated with body mass index.[98] Furthermore, recent studies have emphasized the dependence of the human gut microbial community on diet.[99–101]

Humans appear to be representative of other omnivorous mammals. Turnbaugh described a recent study of humans and 59 other mammalian species from two zoos and the wild. The study results emphasized the broad similarities among the gut microbiota of species spanning the mammalian phylogeny, gut physiologies, and diets.[102] Microbial communities did not group according to habitat (wild versus captive). However, the samples did group based on diet (carnivorous, omnivorous, or herbivorous), with the notable exception of pandas and bears that grouped according to host phylogeny. Interestingly, an analysis of the gene content of these communities revealed that the representation of functional groups of genes correlates with community structure (measured by the 16S rRNA gene sequence) and that carnivore and herbivore microbiota have a greater relative abundance of genes for amino acid catabolism and biosynthesis, respectively.[103]

A major challenge moving forward, Turnbaugh suggested, will be to interface with other fields of study relevant to the gut microbiomes of humans and other mammals. Specifically, what lessons from macro-ecology, wild animal ecology, immunology, and/or parasitology can be applied to studies of the gut microbiome? Can detailed surveys of wild animal populations be used to interpret and guide ongoing studies of the human microbiome? Are the gut microbiomes of laboratory mice representative of their natural colonization, or have they diverged from their wild counterparts? What role do parasites or other small eukaryotes play in shaping the bacterial community and its interactions with the host? And what role does host biogeography play in shaping microbial communities? Together, Turnbaugh concluded, these studies stand to address fundamental biological questions, while also leading toward a promising future of microbiome-based therapies and diagnostics.

Conclusion

The wild immunology symposium concluded with a wide-ranging discussion of the lessons from wild immunology and the next steps for the field. Many of the conclusions immunologists have drawn from the clinic and the laboratory have been reassuringly validated by studies on wild populations, in particular the link between immunological polymorphisms and infectious agents, providing further elegant examples of how pathogens may be maintaining genetic variants in many different host species. The striking parallels between autoimmune propensities in humans, mice, and sheep give further credence to the idea that studies on free-living nonhumans can provide not only a test bed, but a mirror to reveal the relative significance of the many parameters that are thought to regulate optimal immune function. In addition, laboratory-based immunology provides important mechanistic insight that can be incorporated into wild studies and results that need to be tested in natural systems, where their relevance to fitness and health can be investigated. Research in wild immunology will become increasingly valuable as we further investigate the key factors in genetics, ecology, and infection that influence immune status in the real world.

Acknowledgments

The meeting was funded by the Wellcome Trust–funded Centre for Immunity, Infection and Evolution (CIIE), and hosted at the University of Edinburgh. S.A.B. and A.B.P. were funded by CIIE research fellowships. We would also like to thank Daniel Peterson for his contribution to the original, but postponed, Wild Immunology Symposium.

Conflicts of interest

The authors declare no conflicts of interest.

References

1. Pedersen, A.B. & S.A. Babayan. 2011. Wild immunology. *Mol. Ecol.* **20:** 872–880.
2. Norris, K. 2000. Ecological immunology: life history trade-offs and immune defense in birds. *Behav. Ecol.* **11:** 19–26.
3. Schmid-Hempel, P. 2003. Variation in immune defence as a question of evolutionary ecology. *Proc. Biol. Sci.* **270:** 357–366.
4. Nunn, C.L., J.L. Gittleman & J. Antonovics. 2000. Promiscuity and the primate immune system. *Science* **290:** 1168–1170.
5. Møller, A.P. & J. Erritzøe. 2000. Predation against Birds with Low Immunocompetence. *Oecologia* **122:** 500–504.
6. Graham, A.L., A.D. Hayward, K.A. Watt, *et al.* 2010. Fitness correlates of heritable variation in antibody responsiveness in a wild mammal. *Science* **330:** 662–665.
7. Raberg, L., D. Sim & A.F. Read. 2007. Disentangling genetic variation for resistance and tolerance to infectious diseases in animals. *Science* **318:** 812–814.
8. Raberg, L., A.L. Graham & A.F. Read. 2009. Decomposing health: tolerance and resistance to parasites in animals. *Philos. Trans. R. Soc. Lond. B Biol. Sci.* **364:** 37–49.
9. Graham, A.L., D.M. Shuker, L.C. Pollitt, *et al.* 2011. Fitness consequences of immune responses: strengthening the empirical framework for ecoimmunology. *Funct. Ecol.* **25:** 5–17.
10. Bluestone, J.A. 2011. Mechanisms of tolerance. *Immunol. Rev.* **241:** 5–19.
11. Stewart, G.R., M. Boussinesq, T. Coulson, *et al.* 1999. Onchocerciasis modulates the immune response to mycobacterial antigens. *Clin. Exp. Immunol.* **117:** 517–523.
12. Bradley, J.E. & J.A. Jackson. 2008. Measuring immune system variation to help understand host-pathogen community dynamics. *Parasitology* **135:** 807–823.
13. Jackson, J.A., I.M. Friberg, L. Bolch, *et al.* 2009. Immunomodulatory parasites and toll-like receptor-mediated tumour necrosis factor alpha responsiveness in wild mammals. *BMC Biol.* **7:** 16.
14. Mazzoni, A. & D.M. Segal. 2004. Controlling the Toll road to dendritic cell polarization. *J Leukoc Biol* **75:** 721–730.
15. Lambin, X., D.A. Elston, S.J. Petty, *et al.* 1998. Spatial asynchrony and periodic travelling waves in cyclic populations of field voles. *Proc. Biol. Sci.* **265:** 1491–1496.
16. Burthe, S., S. Telfer, X. Lambin, *et al.* 2006. Cowpox virus infection in natural field vole Microtus agrestis populations: delayed density dependence and individual risk. *J. Anim. Ecol.* **75:** 1416–1425.
17. Smith, A., S. Telfer, S. Burthe, *et al.* 2005. Trypanosomes, fleas and field voles: ecological dynamics of a host-vector parasite interaction. *Parasitology* **131:** 355.
18. Telfer, S., M. Begon, M. Bennett, *et al.* 2007. Contrasting dynamics of *Bartonella spp.* in cyclic field vole populations: the impact of vector and host dynamics. *Parasitology* **134:** 413–425.
19. Bown, K.J. 2009. Delineating Anaplasma phagocytophilum Ecotypes in Coexisting, Discrete Enzootic Cycles. *Emerg. Infect. Dis.* **15:** 1948–1954.
20. Gebert, S.F. Helminth dynamics in a cyclic population of field voles. Ph.D Thesis.
21. Jackson, J.A., M. Begon, R. Birtles, *et al.* 2011. The analysis of immunological profiles in wild animals: a case study on immunodynamics in the field vole, *Microtus agrestis. Mol. Ecol.* **20:** 893–909.
22. Viney, M.E., E.M. Riley & K.L. Buchanan. 2005. Optimal immune responses: immunocompetence revisited. *Trends Ecol. Evol.* **20:** 665–669.
23. Abolins, S.R., M.J. Pocock, J.C. Hafalla, *et al.* 2011. Measures of immune function of wild mice, *Mus musculus. Mol. Ecol.* **20:** 881–892.
24. Lochmiller, R.L., M.R. Vestey & S.T. McMurry. 1993. Phenotypic variation in lymphoproliferative responsiveness to mitogenic stimulation in cotton rats. *J. Mammal.* **74:** 189–197.
25. Anderson, R.M., R.M. May & B. Anderson. 1992. *Infectious Diseases of Humans: Dynamics and Control (Oxford Science Publications).* Oxford University Press. New York.
26. Faulkner, H., J. Turner, J. Kamgno, *et al.* 2002. Age- and infection intensity-dependent cytokine and antibody production in human trichuriasis: the importance of IgE. *J. Infect. Dis.* **185:** 665–672.
27. Jackson, J.A., J.D. Turner, L. Rentoul, *et al.* 2004. Cytokine response profiles predict species-specific infection patterns in human GI nematodes. *Int. J. Parasitol.* **34:** 1237–1244.
28. Mackinnon, M.J., T.W. Mwangi, R.W. Snow, *et al.* 2005. Heritability of malaria in Africa. *PLoS Med.* **2:** e340.
29. Quinnell, R.J., R.L. Pullan, L.P. Breitling, *et al.* 2010. Genetic and household determinants of predisposition to human hookworm infection in a Brazilian community. *J. Infect. Dis.* **202:** 954–961.
30. Paterson, S. 1998. Major histocompatibility complex variation associated with juvenile survival and parasite resistance in a large unmanaged ungulate population (*Ovis aries L.*). *Proc. Natl. Acad. Sci. USA* **95:** 3714–3719.
31. Piertney, S.B. & M.K. Oliver. 2006. The evolutionary ecology of the major histocompatibility complex. *Heredity* **96:** 7–21.
32. Jepson, A., W. Banya, F. Sisay-Joof, *et al.* 1997. Quantification of the relative contribution of major histocompatibility complex (MHC) and non-MHC genes to human immune responses to foreign antigens. *Infect. Immun.* **65:** 872–876.
33. Jallow, M., Y.Y. Teo, K.S. Small, *et al.* 2009. Genome-wide and fine-resolution association analysis of malaria in West Africa. *Nat. Genet.* **41:** 657–665.
34. Hill, A.V. 1998. The immunogenetics of human infectious diseases. *Annu. Rev. Immunol.* **16:** 593–617.
35. Ollier, W.E. 2004. Cytokine genes and disease susceptibility. *Cytokine* **28:** 174–178.
36. Fumagalli, M., U. Pozzoli, R. Cagliani, *et al.* 2009. Parasites represent a major selective force for interleukin genes and shape the genetic predisposition to autoimmune conditions. *J. Exp. Med.* **206:** 1395–1408.
37. Snapper, C.M., F.D. Finkelman, D. Stefany, *et al.* 1988. IL-4 induces co-expression of intrinsic membrane IgG1 and IgE by murine B cells stimulated with lipopolysaccharide. *J. Immunol.* **141:** 489–498.
38. Snapper, C.M., C. Peschel & W.E. Paul. 1988. IFN-gamma stimulates IgG2a secretion by murine B cells stimulated

with bacterial lipopolysaccharide. *J. Immunol.* **140**: 2121–2127.

39. Clutton-Brock, T.H. & J.M. Pemberton. 2004. *Soay Sheep: Dynamics and Selection in an Island Population.* Cambridge University Press.

40. Craig, B.H., L.J. Tempest, J.G. Pilkington, *et al.* 2008. Metazoan-protozoan parasite co-infections and host body weight in St Kilda Soay sheep. *Parasitology* **135**: 433–441.

41. Arbuckle, M.R., M.T. McClain, M.V. Rubertone, *et al.* 2003. Development of autoantibodies before the clinical onset of systemic lupus erythematosus. *N. Engl. J. Med.* **349**: 1526–1533.

42. Lleo, A., P. Invernizzi, B. Gao, *et al.* 2010. Definition of human autoimmunity–autoantibodies versus autoimmune disease. *Autoimmun. Rev.* **9**: A259–66.

43. Lipsky, P.E. 2001. Systemic lupus erythematosus: an autoimmune disease of B cell hyperactivity. *Nat. Immunol.* **2**(9): 764–766.

44. Sheldon, B.C. & S. Verhulst. 1996. Ecological immunology: costly parasite defences and trade-offs in evolutionary ecology. *Trends Ecol. Evol.* **11**: 317–321.

45. Clatworthy, M.R., L. Willcocks, B. Urban, *et al.* 2007. Systemic lupus erythematosus-associated defects in the inhibitory receptor FcgammaRIIb reduce susceptibility to malaria. *Proc. Natl. Acad. Sci. USA* **104**: 7169–7174.

46. McNeilly, T.N., E. Devaney & J.B. Matthews. 2009. *Teladorsagia circumcincta* in the sheep abomasum: defining the role of dendritic cells in T cell regulation and protective immunity. *Parasite Immunol.* **31**: 347–356.

47. Kaufman, J. 2010. The Immune Response to Infection. In *The Immune Response to Infection*, 1 ed. S.H.E. Kaufmann, B.T. Rouse & D.L. Sacks, eds. ASM Press. Washington, DC.

48. Kaufman, J. 2008. The Avian MHC. In *Avian Immunology*, 1st ed. F. Davison, B. Kaspers & K.A. Schat, Eds. Elsevier, p. 159–181.

49. Wallny, H.J., D. Avila, L.G. Hunt, *et al.* 2006. Peptide motifs of the single dominantly expressed class I molecule explain the striking MHC-determined response to Rous sarcoma virus in chickens. *Proc. Natl. Acad. Sci. USA* **103**: 1434–1439.

50. Worley, K., M. Gillingham, P. Jensen, *et al.* 2008. Single locus typing of MHC class I and class II B loci in a population of red jungle fowl. *Immunogenetics* **60**: 233–247.

51. Walker, B.A., L.G. Hunt, A.K. Sowa, *et al.* 2011. The dominantly expressed class I molecule of the chicken MHC is explained by coevolution with the polymorphic peptide transporter (TAP) genes. *Proc. Natl. Acad. Sci. USA* **108**: 8396–8401.

52. Kaufman, J., S. Milne, T.W. Göbel, *et al.* 1999. The chicken B locus is a minimal essential major histocompatibility complex. *Nature* **401**: 923–925.

53. Rogers, S.L., T.W. Göbel, B.C. Viertlboeck, *et al.* 2005. Characterization of the chicken C-type lectin-like receptors B-NK and B-lec suggests that the NK complex and the MHC share a common ancestral region. *J. Immunol.* **174**: 3475–3483.

54. Rogers, S.L., B.C. Viertlboeck, T.W. Gobel, *et al.* 2008. Avian NK activities, cells and receptors. *Semin. Immunol.* **20**: 353–360.

55. Kaufman, J. 1999. Co-evolving genes in MHC haplotypes: the "rule" for nonmammalian vertebrates? *Immunogenetics* **50**: 228–236.

56. Salomonsen, J., M.R. Sorensen, D.A. Marston, *et al.* 2005. Two CD1 genes map to the chicken MHC, indicating that CD1 genes are ancient and likely to have been present in the primordial MHC. *Proc. Natl. Acad. Sci. USA* **102**: 8668–8673.

57. Sestak, A.L., B.G. Fürnrohr, J.B. Harley, *et al.* 2011. The genetics of systemic lupus erythematosus and implications for targeted therapy. *Ann. Rheum. Dis.* **70 Suppl 1**: i37–43.

58. Niederer, H.A., L.C. Willcocks, T.F. Rayner, *et al.* 2010. Copy number, linkage disequilibrium and disease association in the FCGR locus. *Hum. Mol. Genet.* **19**: 3282–3294.

59. Smith, K.G.C. & M.R. Clatworthy. 2010. FcgammaRIIB in autoimmunity and infection: evolutionary and therapeutic implications. *Nat. Rev. Immunol.* **10**: 328–343.

60. Ravetch, J.V. & L.L. Lanier. 2000. Immune inhibitory receptors. *Science* **290**: 84–89.

61. Brownlie, R.J., K.E. Lawlor, H.A. Niederer, *et al.* 2008. Distinct cell-specific control of autoimmunity and infection by FcgammaRIIb. *J. Exp. Med.* **205**: 883–895.

62. Clatworthy, M.R. & K.G.C. Smith. 2004. FcgammaRIIb balances efficient pathogen clearance and the cytokine-mediated consequences of sepsis. *J. Exp. Med.* **199**: 717–723.

63. Pritchard, N.R., A.J. Cutler, S. Uribe, *et al.* 2000. Autoimmune-prone mice share a promoter haplotype associated with reduced expression and function of the Fc receptor FcgammaRII. *Curr. Biol.* **10**: 227–230.

64. Jiang, Y., S. Hirose, M. Abe, *et al.* 2000. Polymorphisms in IgG Fc receptor IIB regulatory regions associated with autoimmune susceptibility. *Immunogenetics* **51**: 429–435.

65. Kyogoku, C., H.M. Dijstelbloem, N. Tsuchiya, *et al.* 2002. Fcgamma receptor gene polymorphisms in Japanese patients with systemic lupus erythematosus: contribution of FCGR2B to genetic susceptibility. *Arthritis Rheum.* **46**: 1242–1254.

66. Floto, R.A., M.R. Clatworthy, K.R. Heilbronn, *et al.* 2005. Loss of function of a lupus-associated FcgammaRIIb polymorphism through exclusion from lipid rafts. *Nat. Med.* **11**: 1056–1058.

67. Kono, H., C. Kyogoku, T. Suzuki, *et al.* 2005. FcgammaRIIB Ile232Thr transmembrane polymorphism associated with human systemic lupus erythematosus decreases affinity to lipid rafts and attenuates inhibitory effects on B cell receptor signaling. *Hum. Mol. Genet.* **14**: 2881–2892.

68. Waisberg, M., T. Tarasenko, B.K. Vickers, *et al.* 2011. Genetic susceptibility to systemic lupus erythematosus protects against cerebral malaria in mice. *Proc. Natl. Acad. Sci. USA* **108**: 1122–1127.

69. Willcocks, L.C., E.J. Carr, H.A. Niederer, *et al.* 2010. A defunctioning polymorphism in FCGR2B is associated with protection against malaria but susceptibility to systemic lupus erythematosus. *Proc. Natl. Acad. Sci. USA* **107**: 7881–7885.

70. Fahmy, M.A.M. 1954. An investigation on the life cycle of *Trichuris muris*. *Parasitology* **44**: 50.

71. Cliffe, L.J. & R.K. Grencis. 2004. The *Trichuris muris* system: a paradigm of resistance and susceptibility to intestinal nematode infection. *Adv. Parasitol.* **57:** 255–307.

72. Wakelin, D. 1973. The stimulation of immunity to Trichuris muris in mice exposed to low-level infections. *Parasitology* **66:** 181.

73. Behnke, J.M. & D. Wakelin. 1973. The survival of *Trichuris muris* in wild populations of its natural hosts. *Parasitology* **67:** 157–164.

74. Bancroft, A.J., K.J. Else, N.E. Humphreys, *et al.* 2001. The effect of challenge and trickle *Trichuris muris* infections on the polarisation of the immune response. *Int. J. Parasitol.* **31:** 1627–1637.

75. Smith, K., K.D. McCoy & A.J. Macpherson. 2007. Use of axenic animals in studying the adaptation of mammals to their commensal intestinal microbiota. *Semin. Immunol.* **19:** 59–69.

76. Atarashi, K., T. Tanoue, T. Shima, *et al.* 2011. Induction of colonic regulatory T cells by indigenous *Clostridium* species. *Science* **331:** 337–341.

77. Gaboriau-Routhiau, V., S. Rakotobe, E. Lecuyer, *et al.* 2009. The key role of segmented filamentous bacteria in the coordinated maturation of gut helper T cell responses. *Immunity* **31:** 677–689.

78. Ivanov, I.I., K. Atarashi, N. Manel, *et al.* 2009. Induction of intestinal Th17 cells by segmented filamentous bacteria. *Cell* **139:** 485–498.

79. Round, J.L. & S.K. Mazmanian. 2009. The gut microbiota shapes intestinal immune responses during health and disease. *Nat. Rev. Immunol.* **9:** 313–323.

80. Geuking, M.B., J. Cahenzli, M.A. Lawson, *et al.* 2011. Intestinal bacterial colonization induces mutualistic regulatory T cell responses. *Immunity* **34:** 794–806.

81. Turnbaugh, P.J., R.E. Ley, M. Hamady, *et al.* 2007. The human microbiome project. *Nature* **449:** 804–810.

82. Vijay-Kumar, M., J.D. Aitken, F.A. Carvalho, *et al.* 2010. Metabolic syndrome and altered gut microbiota in mice lacking Toll-like receptor 5. *Science* **328:** 228–231.

83. Wen, L., R.E. Ley, P.Y. Volchkov, *et al.* 2008. Innate immunity and intestinal microbiota in the development of Type 1 diabetes. *Nature* **455:** 1109–1113.

84. Peterson, D.A. & P.J. Turnbaugh. 2010. A microbe-dependent viral key to Crohn's box. *Sci. Transl. Med.* **2:** 43ps39.

85. Khoruts, A., J. Dicksved, J.K. Jansson, *et al.* 2010. Changes in the composition of the human fecal microbiome after bacteriotherapy for recurrent *Clostridium difficile*-associated diarrhea. *J. Clin. Gastroenterol.* **44:** 354–360.

86. Wang, Z., E. Klipfell, B.J. Bennett, *et al.* 2011. Gut flora metabolism of phosphatidylcholine promotes cardiovascular disease. *Nature* **472:** 57–63.

87. Clayton, T.A., D. Baker, J.C. Lindon, *et al.* 2009. Pharmacometabonomic identification of a significant host-microbiome metabolic interaction affecting human drug metabolism. *Proc. Natl. Acad. Sci. USA* **106:** 14728–14733.

88. Wallace, B.D., H. Wang, K.T. Lane, *et al.* 2010. Alleviating cancer drug toxicity by inhibiting a bacterial enzyme. *Science* **330:** 831–835.

89. Dethlefsen, L., S. Huse, M.L. Sogin, *et al.* 2008. The pervasive effects of an antibiotic on the human gut microbiota, as revealed by deep 16S rRNA sequencing. *PLoS Biol.* **6:** e280.

90. Turnbaugh, P.J., V.K. Ridaura, J.J. Faith, *et al.* 2009. The effect of diet on the human gut microbiome: a metagenomic analysis in humanized gnotobiotic mice. *Sci. Transl. Med.* **1:** 6ra14.

91. Sonnenburg, J.L., C.T. Chen & J.I. Gordon. 2006. Genomic and metabolic studies of the impact of probiotics on a model gut symbiont and host. *PLoS Biol.* **4:** e413.

92. Dominguez-Bello, M.G., E.K. Costello, M. Contreras, *et al.* 2010. Delivery mode shapes the acquisition and structure of the initial microbiota across multiple body habitats in newborns. *Proc. Natl. Acad. Sci. USA* **107:** 11971–11975.

93. Turnbaugh, P.J., F. Backhed, L. Fulton, *et al.* 2008. Diet-induced obesity is linked to marked but reversible alterations in the mouse distal gut microbiome. *Cell Host Microbe* **3:** 213–223.

94. Turnbaugh, P.J., R.E. Ley, M.A. Mahowald, *et al.* 2006. An obesity-associated gut microbiome with increased capacity for energy harvest. *Nature* **444:** 1027–1031.

95. Turnbaugh, P.J., M. Hamady, T. Yatsunenko, *et al.* 2009. A core gut microbiome in obese and lean twins. *Nature* **457:** 480–484.

96. Ley, R.E., P.J. Turnbaugh, S. Klein, *et al.* 2006. Microbial ecology: human gut microbes associated with obesity. *Nature* **444:** 1022–1023.

97. Ley, R.E. 2010. Obesity and the human microbiome. *Curr. Opin. Gastroenterol.* **26:** 5–11.

98. Arumugam, M., J. Raes, E. Pelletier, *et al.* 2011. Enterotypes of the human gut microbiome. *Nature* **473:** 174–180.

99. De Filippo, C., D. Cavalieri, M. Di Paola, *et al.* 2010. Impact of diet in shaping gut microbiota revealed by a comparative study in children from Europe and rural Africa. *Proc. Natl. Acad. Sci. USA* **107:** 14691–14696.

100. Walker, A.W., J. Ince, S.H. Duncan, *et al.* 2011. Dominant and diet-responsive groups of bacteria within the human colonic microbiota. *ISME J.* **5:** 220–230.

101. Jumpertz, R., D.S. Le, P.J. Turnbaugh, *et al.* 2011. Energy-balance studies reveal associations between gut microbes, caloric load, and nutrient absorption in humans. *Am. J. Clin. Nutr.* **94:** 58–65.

102. Ley, R.E., M. Hamady, C. Lozupone, *et al.* 2008. Evolution of mammals and their gut microbes. *Science* **320:** 1647–1651.

103. Muegge, B.D., J. Kuczynski, D. Knights, *et al.* 2011. Diet drives convergence in gut microbiome functions across mammalian phylogeny and within humans. *Science* **332:** 970–974.

Ann. N.Y. Acad. Sci. ISSN 0077-8923

Advancing drug discovery for schizophrenia

Stephen R. Marder,[1] Bryan Roth,[2] Patrick F. Sullivan,[3] Edward M. Scolnick,[4] Eric J. Nestler,[5] Mark A. Geyer,[6] Daniel R. Welnberger,[7] Maria Karayiorgou,[8,9] Alessandro Guidotti,[10] Jay Gingrich,[9] Schahram Akbarian,[11] Robert W. Buchanan,[12] Jeffrey A. Lieberman,[8,9] P. Jeffrey Conn,[13] Stephen J. Haggarty,[7] Amanda J. Law,[7] Brian Campbell,[15] John H. Krystal,[16] Bita Moghaddam,[17] Akira Saw,[18,19] Marc G. Caron,[20] Susan R. George,[21] John A. Allen,[22] and Michelle Solis[23]

[1]Semel Institute for Neuroscience at the University of California, Los Angeles; and David Geffen School of Medicine, Los Angeles, California. [2]University of North Carolina School of Medicine, Department of Pharmacology, Chapel Hill, North Carolina. [3]University of North Carolina, Department of Genetics, Chapel Hill, North Carolina. [4]Stanley Center for Psychiatric Research, Broad Institute of MIT and Harvard University, Cambridge, Massachusetts. [5]Mount Sinai Medical Center, New York, New York. [6]University of California, San Diego, La Jolla, California. [7]Clinical Brain Disorders Branch, Genes, Cognition, and Psychosis Program, Intramural Research Program, National Institute of Mental Health, National Institutes of Health, Bethesda, Maryland. [8]Department of Psychiatry, Columbia University, New York, New York. [9]New York State Psychiatric Institute, New York, New York. [10]University of Illinois at Chicago, Chicago, Illinois. [11]Research Institute and Program in Bioinformatics and Integrative Biology, University of Massachusetts Medical School, Worcester, Massachusetts. [12]University of Maryland School of Medicine, Baltimore, Maryland. [13]Vanderbilt Program in Drug Discovery, Department of Pharmacology, Vanderbilt University Medical Center, Nashville, Tennessee. [14]Center for Human Genetic Research, Massachusetts General Hospital, Harvard Medical School, Boston, Massachusetts. [15]Pfizer Neuroscience, Groton, Connecticut. [16]Yale University School of Medicine, New Haven, Connecticut. [17]Department of Neuroscience, University of Pittsburgh, Pittsburgh, Pennsylvania. [18]Department of Psychiatry and Behavioral Sciences, Johns Hopkins University School of Medicine, Baltimore, Maryland. [19]Department of Neuroscience, Johns Hopkins University School of Medicine, Baltimore, Maryland. [20]Department of Cell Biology, Duke University Medical Center, Durham, North Carolina. [21]Centre for Addiction and Mental Health, Departments of Medicine and Pharmacology, University of Toronto, Toronto, Ontario, Canada. [22] Pfizer Neuroscience, Groton, Connecticut; and University of North Carolina School of Medicine, Department of Pharmacology, Chapel Hill, North Carolina. [23]Schizophrenia Research Forum

Sponsored by the New York Academy of Sciences and with support from the National Institute of Mental Health, the Life Technologies Foundation, and the Josiah Macy Jr. Foundation, "Advancing Drug Discovery for Schizophrenia" was held March 9–11 at the New York Academy of Sciences in New York City. The meeting, comprising individual talks and panel discussions, highlighted basic, clinical, and translational research approaches, all of which contribute to the overarching goal of enhancing the pharmaceutical armamentarium for treating schizophrenia. This report surveys work by the vanguard of schizophrenia research in such topics as genetic and epigenetic approaches; small molecule therapeutics; and the relationships between target genes, neuronal function, and symptoms of schizophrenia.

Keywords: schizophrenia; genetics; GWAS; neuronal function; small molecules; therapeutics

Background and perspectives: keynote lectures

Session chairs: Stephen R. Marder and Bryan Roth

Though pioneering pharmacotherapy for the treatment of schizophrenia occurred in the 1950s with the introduction of antipsychotics, there have been decidedly few major innovations in the intervening years. Indeed, the current classes of drugs available largely involve the same mechanisms of action and the same neurobiological targets. Compounding the lack of progress has been the systemic pull-back of the pharmaceutical industry from research and development of new treatments for schizophrenia.

With the advent of successful methods for identifying anomalies at the genetic level, schizophrenia and other neuropsychiatric disorders stand to

doi: 10.1111/j.1749-6632.2011.06216.x

benefit immensely from incorporating these approaches and broadening our understanding of the pathogenesis of these illnesses.

"Advancing Drug Discovery for Schizophrenia" convened with an opening session of keynote lectures moderated by session chairs Stephen R. Marder (University of California, Los Angeles) and Bryan Roth (University of North Carolina School of Medicine). Marder called for a change in the way schizophrenia treatment has previously been conceptualized by moving away from focusing only on treating symptoms in individuals who are in the acute phase of the illness. Innovative drugs may emerge from improving our understanding of the pathologic processes that occur early in the development of the illness, thus providing strategies for altering the course of the disorder. Importantly, genetic and epigenetic approaches present a very promising means by which to accomplish this goal.

Genetic and epigenetic approaches to studying schizophrenia

Patrick F. Sullivan (University of North Carolina at Chapel Hill) introduced the advantages of genetic approaches to schizophrenia by listing several benefits: the temporality of exposure to genetic risk factor preceding disease can be reasonably assumed; attempts at unbiased searches are possible; and, in the past two years, there has been substantial progress in uncovering the genetic basis of schizophrenia. He suggested that the purpose of employing genetic techniques is to look both for genetic variance that confers risk or prevention and for whether multiple genes work in concert to form pathways to an illness. To illustrate this point, he offered the analogy of a match to a forest fire—genetic studies seek to identify the initiating event in a disorder, or in other words, the "match" that sparks the fire.

This line of work has begun to make progress.[1,2] Sullivan noted that there are novel and robust findings that include common variation in the major histocompatibility complex, the region encoding the microRNA miR-137, and four predicted targets of miR-137. Copy number variants that are rare but have high penetrance with robust results include regions on 1q21.1, 15q13.3, 16p11.2, 22q11.2, and *NRX1*. At this point, up to 15 possible genetic loci for schizophrenia have been identified. Further, Sullivan stressed that the accumulated data suggest that

Table 1. Genetic characteristic of SCZ

SCZ is highly polygenic
• Highly significant, replicated, neg control
• Withstood attempts at falsification
• (New) Not due to pop stratification
• Accounts for one-third of variance in liability to SCZ (two methods), half heritability
• Genetic overlap between SCZ and BIP

schizophrenia is relatively highly polygenic, which may be a general feature of complex traits. For example, various genes may actually code at the symptom level versus the disorder level (Table 1).

The results to date are notable, Sullivan underscored, in that the novelty of findings correspond relatively poorly to prior ideas about the etiology of schizophrenia. Empirical data suggest that both common and rare variations are involved in the etiology of schizophrenia. Rare variant–only models have been evaluated and are not consistent with the data. Sullivan suggested that with genome-wide association studies (GWAS), we can, however, estimate that increasing sample sizes are likely to produce more significant loci and considerably better pathway analyses. He likened GWAS to a powerful centrifuge that allows for the separation of specific genes. Most heartening, he suggested, is the likelihood that the discovery profile for schizophrenia has a linear trajectory—increasingly larger samples will lead to an increasing number of loci identified.

Genetic architecture of psychotic illness

Edward M. Scolnick (Broad Institute of MIT and Harvard University) began by suggesting the great importance of unraveling the full genetic architecture of schizophrenia and bipolar disorder as the critical step in establishing their underlying neurobiology. While the overarching theme of the conference pointed to a general stagnation in the field of schizophrenia treatment, Scolnick lauded the significant strides made recently. In other words, the field has become "unstuck." This movement could also be attributed to the application of three new technologies: human genetics, optogenetic methods to manipulate specific circuits in the brain, and induced pluripotent stem cells.

Scolnick noted the apparent disparity between developments in schizophrenia versus cancer treatments. Bolstering this swell in cancer drug development is the increased understanding of the genetic and molecular aspects of various forms of cancer, something that, given the limited knowledge of its genetic basis, is not yet possible for schizophrenia.

In the past three years, however, using detailed maps of the human genome and doing association studies, a number of risk genes for schizophrenia and bipolar illness have been identified. CNVs (copy number variants) and genomic associations have been identified with a high degree of statistical certainty; some of the CNVs are *de novo* and some are inherited. Scolnick referred to characterizing diseases by the frequency of variants in addition to penetrance. For example, Mendelian diseases possess high penetrance and are deterministic of the development of the disease. However, many diseases posses a mixture of high- and low-frequency variation, indicated Scolnick. Increasingly, GWAS methodology is being used to investigate the penetrance of diseases and is appropriate for application to schizophrenia. Pathways involving synaptic proteins, the WNT signaling pathway, as well as miR-137 have been identified, which now provide clues about the biology of some illnesses. In addition, it has become clear that schizophrenia and autism may share some risk genes in common. Though the outcomes are very different, it may be possible, Scolnick suggested, that causative DNA variants may someday be used diagnostically.

Scolnick concluded by posing the idea that the full-scale application of genomics is now limited only by money and time to unravel the full genetic architecture of these illnesses. Further, given the burden of these illnesses, he reasoned that there is "no excuse" for NIH and NIMH funds not to be made available to perform the needed genetic studies and to finally enlighten their etiologies. He continued by saying that until we understand fully the genes that cause these illnesses the field will remain a "backwater of medicine." Funding for neuroscience is certainly needed to understand their pathophysiology, but without an understanding of the genetic bases, he cautioned, scientists will continue to guess at what to work on to elucidate the illness and discover new targets for treatment.

What can we learn from animal models of schizophrenia?

Applying genetic methodologies to psychiatric disorders presents unique challenges, suggested Eric J. Nestler (Mount Sinai Medical Center). These include that they are categorically defined by behavioral abnormalities and that the genetic risks involved are likely to be a combination of multiple genetic variations. Thus, progress in understanding the etiology and pathophysiology of schizophrenia has been frustratingly slow, as has been the discovery of significant, novel therapeutic mechanisms. The exceedingly challenging neurobiology of higher brain function and the ethical and practical difficulties of examining the living human brain have no doubt contributed to this relatively slow progress. Another important factor is the extremely challenging nature of modeling schizophrenia in laboratory animals.[3] This is due to the inaccessibility, in animals, of many of the key symptoms of schizophrenia, the subjective nature of the symptoms, the lack of objective biomarkers, and the early state of knowledge of the underlying genetics and neurobiology of the syndrome.

What we have learned from genetic studies of schizophrenia thus far, Nestler argued, is especially complicating vis-à-vis animal models, because most genes of relatively strong effect implicated in schizophrenia are also implicated in bipolar disorder and even autism. Moreover, Nestler proposed that what can be learned from placing genes of very small effect into laboratory animals is itself questionable. Indeed, Nestler continued by saying that most available animal models of schizophrenia have significant limitations, ranging from weak validation to poor predictive power for drug efficacy in humans.

Illustrating the complications that arise from translating models of human disorders in, for example, murine models, Nestler explored difficulties for establishing construct validity. For example, though genes may be associated with schizophrenia, abnormalities in these specific genes may cause schizophrenia but may also cause other disorders. Moreover, there is difficulty in making animal models as complicated as human models, and this too may require experimentally manipulating multiple genes.

Nestler also discussed two other forms of validity in respect to the limitations of animal models.

Perhaps the most obvious limitation is the lack of face validity—there are simply features of psychiatric disorders that are not accessible in animals. Second, Nestler referenced the lack of predictive validity of animal models. Not only are there no new medications to test for efficacy in animals, but we must also interpret any response to drugs with great caution. In contrast, there may be better face validity for some of the negative symptoms associated with schizophrenia (e.g., social withdrawal and anhedonia), which have been modeled successfully, but may not be unique to this illness.

Concluding his talk, Nestler remained optimistic that, with time, the genes that confer risk or resilience in mental illness will be determined, citing advances in autism as support. The generation of convincing and useful animal models of schizophrenia thus represents a major set of challenges that will not have easy answers. On the other hand, given current limitations of clinical studies, Nestler concluded that it is hard to imagine significant progress in schizophrenia pathophysiology or therapeutics without good animal models.

Session II: Genetic and epigenetic approaches

Session chair: Mark A. Geyer

A novel therapeutic target for psychosis
The discovery of genes associated with risk for schizophrenia holds promise for identifying therapeutic targets based on etiopathogenesis. Daniel R. Weinberger (National Institute of Mental Health) reported recent work from his lab documenting an association with schizophrenia to the hERG family potassium channel gene, KCNH2.[4] Risk-associated variants predict expression of a novel brain and primate specific isoform, KCNH2 3.1, which both has unique physiological properties and is upregulated in schizophrenia prefrontal and hippocampal cortices. Because many antipsychotic drugs bind to KCNH2 channels, accounting in some cases for QT prolongation, and atypical agents such as clozapine are particularly potent KCNH2 inhibitors, Weinberger and coworkers hypothesized that activity at KCNH2 channels may be a therapeutic mechanism of antipsychotic drugs. As a result, targeting the 3.1 isoform not expressed in the heart would have a particularly favorable therapeutic-to-toxic ratio. Weinberger stated, "We

have tested the first hypothesis in two independent samples of patients who underwent placebo controlled studies of antipsychotic therapy, including an inpatient sample from the CBDB/NIMH experimental therapeutics unit ($n = 59$) and Caucasian subjects treated with olanzapine as part of the CATIE trial from whom we also had drug clearance data (total $N = 89$)." In both groups, Weinberger and colleagues found that the KCNH2 genotype predicted response to treatment (e.g., positive symptoms $P < 0.0007$, thought disturbance $P < 0.0004$ in the CBDB sample, positive symptoms $P < 0.04$ in CATIE). Currently, Weinberger's group is testing the second hypothesis in human and model cell lines that over-express KNCH2 3.1 and also in a mouse model that overexpresses KCNH2 3.1 in cortex and shows predicted abnormalities in K^+ channel kinetics and in working memory.

Linking DNA structural variation to brain dysfunction and schizophrenia
Structural variation in the genome in the form of copy-number variants (CNVs; submicroscopic gains or losses of segments of DNA) has now been identified as a major contributing genetic factor in schizophrenia.

Maria Karayiorgou (Columbia University) discussed the importance of examining low-frequency alleles with moderate effects, as they are easier to characterize and potentially more likely to contribute to the development of the disorder. Further, exploring the genetic variation in the landscape *de novo* may have great utility in delineating sporadic versus inherited types of schizophrenia. Importantly, Karayiorgou suggested that inherited CNVs are more disrupting than those that are *de novo*.

Karayiorgou focused on one such CNV on chromosome 22q11.2. Recent studies are beginning to paint a clearer and more consistent picture of the impairments in psychological and cognitive competencies that are associated with microdeletions in chromosome 22q11.2. In contrast to previous talks stressing the uncertainty of associating schizophrenia risk with any specific gene or CNV, Karayiourgou maintained that 30% of individuals with a deletion on 22q11 develop schizophrenia or schizoaffective disorder. Indeed, parallel studies in humans and animal models have begun to uncover the complex genetic and neural substrates altered by this microdeletlon. For example, Krayiorgou illustrated studies that have begun to identify CNVs that may explicate some of the working memory-related processing deficits seen in schizophrenia, because they occur in regions that oversee microRNA, thus affecting prefrontal cortex neuronal synaptic transmission and plasticity. In addition to offering a deeper understanding of the effects of this genetic lesion, Krayiorgou ended her presentation with the hope that these findings may guide analysis of other CNVs associated with psychiatric disorders.

Targeting the neuro-epigenome in the prevention and treatment of psychosis

Alessandro Guidotti (University of Illinois at Chicago) began by stating that the efficacy of current drugs is limited and that, without new developments, there is little hope for progress in the treatment of schizophrenia. An area of research that holds promise is exploring the role of the fundamental interaction between GABAergic interneurons and pyramidal neurons and how this function is compromised in schizophrenia and bipolar disorders. Schizophrenia and bipolar disorder patients show a downregulation of GAD_{67}, reelin, and other genes expressed in telencephalic GABAergic neurons, likely associated with promoter hy-

permethylation mediated by an overexpression of DNA methyltransferase (DNMT).[5] The inhibitory action of DNMT on gene expression may also occur through formation of chromatin repressor complexes that include histone deacetylases (HDAC) or histone methyltransferases, thereby shifting chromatin from a conformation permissive for transcription to one that is repressive. Guidotti has concentrated on a pharmacological strategy to reduce the hypermethylation of GABAergic promoters and to induce a permissive chromatin conformation. This strategy is to administer drugs such as the HDAC inhibitor valproate (VPA), which induces DNA demethylation when administered at doses that facilitate chromatin remodeling.[6]

Studies employing this strategy in mice, as presented by Guidotti, suggest that when combined with VPA, clinically relevant doses of clozapine elicit a synergistic potentiation of VPA-induced GABAergic promoter demethylation (Fig. 1). Olanzapine and quetiapine (two similar atypical antipsychotic drugs) also facilitate chromatin remodeling, but at doses higher than used clinically, whereas haloperidol and risperidone are inactive. Hence, Guidotti suggested the synergistic potentiation of VPA's action on chromatin remodeling by clozapine appears to be a unique property of the dibenzepines and is independent of their action on catecholamine or

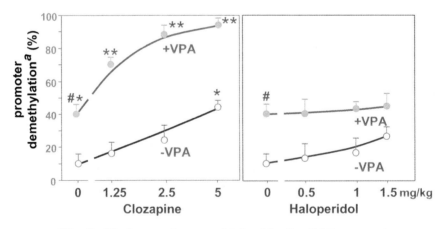

^a*Reelin*. Similar results were obtained for the GAD_{67} promoter

Figure 1. Clozapine but not haloperidol, alone or in combination with VPA, induces *reelin* promoter demethylation. Mice were pretreated for seven days with methionine to hypermethylate reelin and GAD_{67} promoters. After methionine withdrawal, various doses of clozapine or haloperidol alone or combined with valproate (VPA) (0.7 mg/kg/sc) were administered twice a day for three days. Open circles denote mice that did not receive VPA. Filled circles denote mice that received VPA. The data represent the mean ± SE of three mice. $P < 0.05$ when VEH + VPA-treated mice were compared with VEH-treated mice; $P < 0.05$ when VEH + clozapine-treated mice were compared to VEH-treated mice; and $P < 0.02$ when VEH + VPA-treated mice were compared with clozapine + VPA-treated mice.[7]

serotonin receptors. Furthermore, the administration of clozapine in conjunction with VPA reverses the downregulation of the GAD_{67} expression induced in mice by seven days of methionine administration.[5]

Guidotti concluded by positing that chromatin remodeling mechanisms may be altered in the brains of schizophrenia and bipolar disorder patients. Thus, by activating DNA demethylation, the association of clozapine or its derivatives with VPA or other more potent and selective HDAC inhibitors could be considered a promising prevention and treatment strategy to normalize the GABAergic promoter hypermethylation and GABAergic gene expression downregulation detected in the postmortem brain of schizophrenia and bipolar patients.

Advanced paternal age alters complex behaviors and brain DNA methylation in offspring

Prompted by epidemiological research that shows advanced paternal age increases risk for schizophrenia in offspring,[a] Jay Gingrich (Columbia University Medical School) discussed potential epigenetic mechanisms behind this effect. With each round of sperm production, epigenetic marks are erased and then reprogrammed—a process that may become degraded with advanced paternal age. To explore this, Gingrich and colleagues compared the offspring from older mouse fathers to those from young mouse fathers. Several behaviors differed between the two, including measures of an open-field ambulatory test, startle responses, and paired pulse inhibition. These mice showed distinct methylation patterns on their genomes at 0.4 % of the methylation sites. At these sites, the offspring of older males were less methylated than those of the younger males in introns, exons, and promoters. These findings suggest that differences inboth the promoter activity and alternative splicing patterns may drive some of the changes observed in the mice. Notably, hotspots of altered DNA methylation could be found in schizophrenia-related genes, though, Gingrich noted, the researchers didn't select the mice for disease-related behaviors.

---

[a]See the SRF hypothesis: http://www.schizophrenia forum.org/for/curr/Malaspina/default.asp

Neuronal epigenome mapping in developing and diseased prefrontal cortex

Alterations in chromatin structure and function, including changes in levels or distribution of histone lysine methylation markings and other epigenetic regulators of gene expression, could affect neuronal signaling in schizophrenia and other major psychiatric disease.[8] However, to date, relatively nothing is known about the regulation of neuronal and other cell type–specific epigenomes in the diseased brain. Schahram Akbarian (University of Massachusetts) opened his talk by pointing out the potential for neuroepigenome mapping approaches to advancing the epigenetic field through focusing on the postmortem human brain. Specifically, he discussed work exploring the genome-wide distribution of trimethylated histone H3K4 (H3K4me3), an epigenetic histone mark associated with actual or potential transcription, in neuronal nuclei collected postmortem from prefrontal cortex (PFC) across a wide age range (0.5–70 years).

Neuronal nuclei from postmortem prefrontal cortex were immunotagged with NeuN antibody, $NeuN^+$ and $NeuN^-$ nuclei were sorted separately via fluorescence-activated "cell" (nuclei) sorting (FACS), and purified mononucleosomes were enriched for H3K4me3 analyzed by massively parallel sequencing using an Illumina Solexa platform.[9] Akbarian's group has identified 16,000–22,000 H3K4me3-enriched regions (peaks), the majority located proximal to (within 2 kb of) the transcription start sites (TSS) of annotated genes. These included signatures specific to neurons as well as signatures specific to individual subjects (Fig. 2). Preliminary analyses presented by Akbarian provided evidence for a large-scale remodeling of histone methylation landscapes in immature PFC neurons during early infancy, with comparatively minor changes during subsequent periods of maturation and aging. Further, he predicted that epigenome mapping in defined cell populations of the diseased human brain, in conjunction with RNA and transcriptome profilings, could reveal important information about gene expression alterations at specific loci. These approaches, he stressed, will also be important to unravel the genetic and epigenetic risk architectures of specific cases and, furthermore, reveal insight about developmental regulation of chromatin structures in specific cell populations.

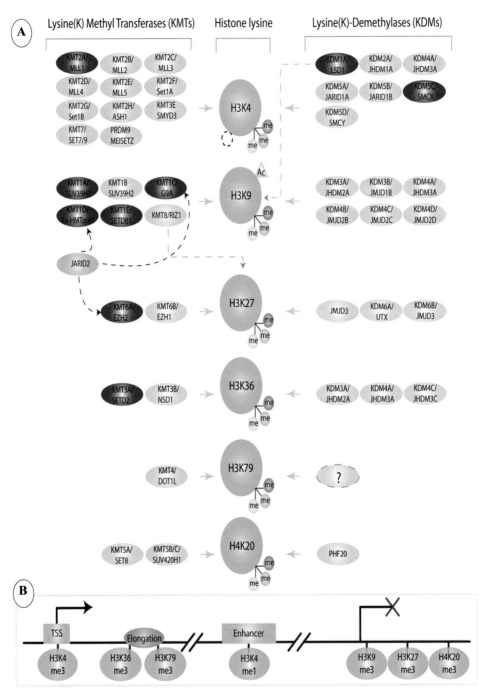

Figure 2. Cell type–specific epigenome mapping in human brain. Left: schematic presentation of the nucleosome as the elementary unit of chromatin, composed of a histone octamer (blue) around which 146 bp of DNA (black) is wrapped. Histones and the genomic DNA are subject to various types of covalent modifications. Right: flow chart starting with extraction of nuclei from postmortem brain tissue for subsequent immunotagging with neuron nuclei–specific marker (NeuN), fluorescence-activated separation and sorting of neuronal and nonneuronal nuclei, and preparation of chromatin with enzyme-based digestion (micrococcal nuclease, MNase) into mononucleosomes. This is followed by immunoprecipitation with site- and modification-specific antihistone antibody (for example, antitrimethyl-histone H3-lysine 4) and massively parallel sequencing. Sequence tags from the immunoprecipitates (ChIP-seq) are then uploaded into the genome browser in order to visualize histone methylation landscapes at selected loci, as shown here for the GAD1/GAD67 promoter.

Session III: New molecular targets and approaches to small molecule therapeutics

Session chair: Robert W. Buchanan

Development of novel antipsychotic drugs: new approaches to old problems

Dopamine D2 receptor antagonism is a unifying property of all antipsychotic drugs in use for schizophrenia. Jeffrey A. Lieberman (Columbia University and New York State Psychiatric Institute) stressed during his talk that, as a result, while these drugs can be effective at ameliorating psychosis, they are largely ineffective at treating negative and cognitive symptoms, can produce serious side effects involving different organ systems, and may even contribute to gray matter volume reduction. The differences between the majority of the current drugs, Lieberman suggested, are primarily due to side-effect profiles, with pronounced side-effects often outweighing the therapeutic potential; additionally, considerable heterogeneity in response to treatment exists.

For years, psychiatric researchers and pharmaceutical companies have sought to develop mechanistically novel antipsychotic drugs based on targets other than dopamine receptors. For example, Lieberman highlighted the considerable amount of work that has been devoted to precedented targets and mechanisms (D-2, D-3 dopamine, and 5-HT2A serotonin antagonism); novel mechanisms (partial agonism, functional selectivity); and novel targets (glutamatergic, cholinergic, GABAergic), including intracellular signaling mechanisms (AKT, GSK). Unfortunately, drugs targeting many of these mechanisms were abandoned before they were tested in clinical trials.

In addition, Lieberman offered the observation that increasing attention is being focused on the complex genetics of schizophrenia as well as the signaling pathways implicated in its pathophysiology. Lieberman's presentation reviewed the limitations of existing therapeutic agents and development strategies, including several of the major genetic findings that have identified signaling pathways representing potential targets for novel pharmacological intervention—for example, genes in the 22qll locus, DISC1, Neuregulin1/ ErbB4, and components of the Akt/GSK-3 pathway.

Lieberman concluded that it may be critical both to consider the necessity of multiple medications that treat different aspects (or domains) of schizophrenia and to renew commitment to research and development of pharmacotherapy for the illness.

Allosteric modulators of metobotropic glutamate receptor 5 for the treatment of schizophrenia

P. Jeffrey Conn (Vanderbilt University) presented work demonstrating the path between hypotheses and actual treatments of schizophrenia. The glutamate/NMDA receptor function hypofunction hypothesis has garnered wide attention because of its significant potential in targeting the range of symptoms presented in schizophrenia, including positive, negative, and cognitive symptoms. Further supporting this hypothesis, a large number of cellular and behavioral studies suggest that selective agonists of the metabotropic glutamate receptors (mGluRS) could provide a novel approach to treatment of schizophrenia.[10] Unfortunately, it has been difficult to develop compounds that act as selective orthosteric agonists of mGluRS and that have properties likely to be suitable for development of therapeutic agents. Conn presented work from his group, which has been successful in implementing a novel approach to activation of these receptors by developing highly selective allosteric potentiators of mGluR5. These compounds do not activate mGluR5 directly, but rather potentiate the response of these receptors to the endogenous agonist. Thus, these allosteric potentiators offer high selectivity for the targeted receptor and provide an exciting new approach to development of novel selective activators of mGluR5 and other G protein–coupled receptor (GPCR) subtypes.[10]

In vivo studies reveal that these compounds have robust effects in animal models, which have been used to predict efficacy of novel antipsychotic agents. Conn showed findings that demonstrate that introduction of these compounds reduces amphetamine-induced motor hyperactivity, a putative animal model of limbic hyperactivity and positive symptoms of schizophrenia. In addition, mGluR5-positive allosteric modulators (PAMs) enhance multiple forms of synaptic plasticity in the central nervous system (CNS) and have cognition-enhancing effects in animal models.[11] These studies, therefore, provide an exciting new approach for discovery of novel, highly selective activators of specific

GPCR subtypes for treatment of schizophrenia and other CNS disorders. In conclusion, Conn suggested that recent advances in the discovery of a broad range of mGluR5 allosteric modulators with different functional profiles will allow the development of a more complete understanding of the properties of individual mGluR5-PAMs with the hope of determining which may be most suitable for development as novel therapeutic agents.

Chemical genomic studies of DISC1/GSK3–mediated signaling in neuropsychiatric disorders

A critical need exists for gaining insight into the underlying etiology and pathophysiology of schizophrenia and other neuropsychiatric disorders in order to advance the development of new types of targeted therapeutic interventions. Surveying recent work, Stephen J. Haggarty (Harvard Medical School, Broad Institute of MIT and Harvard University) presented studies of the Disrupted in Schizophrenia 1 (DISC1) gene, which is disrupted by a balanced chromosomal translocation t(1;11)(q42.1;q14.3) in a single extended Scottish family with a high incidence of schizophrenia, major depression, and bipolar disorder. Studies have revealed a key role

for DISC1-mediated signaling in diverse aspects of neuroplasticity. One of the direct binding targets of DISC1 is the multifunctional serine/threonine kinase GSK3, which is known to play a key role in the regulation of neurogenesis and synaptic function.[12] GSK3 has also been shown to be inhibited *in vivo* directly, and indirectly, by the mood stabilizer lithium and indirectly through the activity of antipsychotics at various GPCRs.[13]

On the basis of these findings, as well as emerging human genetic studies implicating other components of the GSK3 signaling pathway in the etiology of schizophrenia, Haggarty discussed the advent of chemical genomic approaches to identify small molecule probes that target known and novel components of DISC1/GSK3 signaling.

Using pharmacological and viral-mediated gene expression to modulate AKT kinase activity, Haggarty presented work demonstrating a key role for AKT kinase activity in determining the cellular and mood-related behavioral effects of lithium.[13,14] In contrast, selective and direct GSK3 inhibition by an ATP-competitive, highly selective, brain penetrant GSK3 inhibitor (CHIR-99021) was found to bypass the requirement for AKT activation (Fig. 3).[14] In order to try to develop

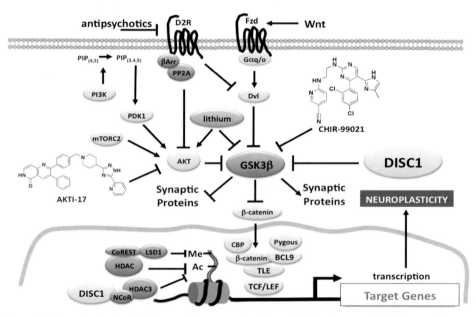

Figure 3. DISC1/GSK3 signaling involved in regulating neuroplasticity and its regulation. While lithium inhibits GSK3 both indirectly and directly, CHIR-99021 is an ATP competitive inhibitor of GSK3 that mimics the cellular and behavioral effects of lithium. AKTI-17 is an allosteric inhibitor of AKT that blocks the ability of lithium to modulate TCF/LEF–mediated transcription and behavior but not the effect of CHIR-99021.

pathway-selective inhibitors of GSK3 and to develop small molecule probes that enable the selective targeting of individual DISC1 domains, Haggarty's group has used small molecule microarray screening to measure the interaction of DlSC1 variants in a high throughput manner.

Finally, Haggarty presented efforts to use induced pluripotent stem cell (iPSC)–derived human neurons for use in high-throughput screening for small molecule probes of DISC1/GSK3 signaling. Haggarty showed findings that demonstrated the ability to identify known and novel small molecules that regulate TCF/LEF–dependent transcription that is under control of GSK3 signaling in iPSC-derived human neurons. Collectively, he noted, through the use of genetically accurate cell models and in using iPSC-derived human neurons, these studies will enable a better understanding of the role of DISC1/GSK3 in regulating signaling pathways implicated in schizophrenia, as well as potentially lead to the identification of new targets and improved treatments.

Targeting the PI3K/AKT pathway: novel therapeutic options for schizophrenia

Neuregulin 1 (NRG1) and ErbB4, critical neurodevelopmental genes, are implicated in genetic risk for schizophrenia, but the biological mechanisms are unknown. Amanda J. Law (National Institute of Mental Health) presented work testing the hypotheses that aberrant PI3K/AKT signaling represents a pathogenic consequence of schizophrenia-associated genetic variation in ErbB4 and that pharmacological manipulation of this pathway may represent a novel therapeutic avenue for the treatment of psychosis. She explained that her group has used an "integrative translational neuroscience approach" to test the hypothesis, incorporating data from patient-derived lymphoblastoid B cells, human brain mRNA expression profiling, clinical genetics, and pharmacological studies in rodents. These results pinpoint a genetically regulated pathway associated with schizophrenia and with ErbB4 genetic risk variation involving upregulation of a PI3K-linked ErbB4 receptor CYT-l and of a specific phosphoinositide 3-kinase (PI3K) subunit. Law presented intracellular signaling data collected in human lymphoblastoid cells showing that NRG1-mediated phosphatidyl-inositol,3,4,5-triphosphate [PI(3,4,5)P3] production (and cell migration) is as-

sociated with ErbB4 risk genotype and PI3K enzyme levels and is significantly impaired in patients with schizophrenia. In the human brain, the association of ErbB4 genotype and PI3K subunit expression was confirmed, and antipsychotic drug administration was found to downregulate expression of the PI3K subunit, implicating the PI3K pathway as a novel therapeutic target. Specific inhibition of the PI3K subunit using a small molecule inhibitor blocks the effects of amphetamine in a mouse pharmacological model of psychosis. Law posited that these studies provide novel insight into how ErbB4 may contribute to the pathophysiology of schizophrenia, reveal a previously unidentified link between NRG1-ErbB4 and PI3K signaling in the disorder, and suggest that a specific PI3K merit further consideration as a molecular target for rationally designed drugs for the treatment of psychiatric disorders.

Evaluation of potent, selective, and peripherally available kynurenine aminotransferase II inhibitors for the treatment of cognitive deficits in schizophrenia

Representing work conducted in the private sector, Brian Campbell (Pfizer) discussed recent work on a naturally occurring NMDA agonist, kynurenic acid (KYNA). KYNA is a biologically active byproduct of tryptophan metabolism that is reported to act as an endogenous antagonist of NMDA receptors and may also interfere with nicotinic α7 receptor function. Since KYNA levels are elevated in schizophrenic patients, it has been hypothesized that this may contribute to the cognitive deficits observed in schizophrenia. Though inhibitors of kynurenine aminotransferase 11 (KAT II), the primary enzyme involved in brain KYNA synthesis, have been proposed as targets to treat symptom domains in schizophrenia, research into this hypothesis has been hampered by lack of potent, selective, and brain penetrant tools. Campbell reported on the discovery and pharmacological characterization of a new class of KAT II inhibitors with nanomolar potency, >1000-fold selectivity over other kynurenine pathway enzymes, and excellent brain penetration. Systemic administration rapidly decreased KYNA by as much as 80% in brain dialysates in a dose-dependent manner. In addition, Campbell detailed the profiling of these compounds through a range of animal models relevant for domains of schizophrenia, which revealed efficacy in models of working memory in both rodents and non-human primates and also in a rat model of attention (i.e., a sustained attention task). Furthermore, Campbell and the group at Pfizer have found activity in an anhedonia model (chronic mild stress), which may suggest therapeutic utility for the negative symptoms of schizophrenia. These compounds are inactive in classical models of antipsychotic activity, suggesting that KAT II inhibitors may be suitable as adjunctive agents, given along with antipsychotics, to treat the cognitive and negative symptoms associated with schizophrenia that are so poorly managed today.

Genes to function to symptoms

Session chair: John H. Krystal

Large-scale neuronal analysis as a functional endophenotype for schizophrenia

Bita Moghaddam (University of Pittsburgh) opened the final session of the meeting. Dissatisfied with how simple circuit versions of the glutamate hypothesis of schizophrenia fail to deliver for the whole brain, she proposed an alternative way of thinking about the disorder. Instead of seeing the main clues we have for schizophrenia—NMDA receptor hypofunction and decreased GAD67, an enzyme that synthesizes GABA—as factors that contribute to the disorder, she suggested that decreased GAD67 may instead reflect a compensation for abnormal levels of glutamate. If markers of disease like these are compensatory, rather than primary, then trying to correct them would give opposite effects than expected. With this in mind, she cited a tightly regulated GABA shunt in mitochondria that recycles GABA molecules from glutamate. Moghaddam suggested that imbalances in GABA and glutamate levels could stem from improper mitochondrial function—something with repercussions for excitatory and inhibitory signaling throughout the brain.

Translational neuroscience for schizophrenia

Seeking to better simulate the molecular processes underway in neurons in disease, Akira Sawa (Johns Hopkins University) reported on the "garden of human neural cells" growing in his lab. He has had some success in maintaining olfactory neurons from humans obtained through a relatively noninvasive nasal biopsy,[15] deriving human neurons from iPSCs,[16] and deriving neurons directly from human skin cells.[17] After two weeks of treatment in media,

these latter neurons, called induced neurons (iN), showed neuron-specific markers and fired action potentials.

As an example of the dividends these approaches may provide, Sawa showed a comparison of mRNA profiles in olfactory neurons between people with schizophrenia and controls. The groups differed in genes related to actin binding, the NF-κB protein complex involved in DNA transcription, intracellular protein transport, and immune and stress responses. Consistent with these changes, olfactory neurons from patients with schizophrenia show abnormalities in association with cytoskeletons and oxidative stress (Table 2).

Behavioral selectivity of β-arrestin: dependent signaling of dopamine D2 receptor in the CNS

The last three speakers discussed whether there might be ways to refine dopamine signaling in the brain to treat schizophrenia more effectively and with fewer side effects. Starting with D2Rs, the target of all antipsychotics, Marc Caron (Duke University) noted that D2Rs engage both a G protein–coupled pathway and one involving the scaffolding protein β-arrestin-2, which in turn activates the AKT/GSK3β signaling independently associated with schizophrenia.

To probe how much this β-arrestin-2 pathway mediates psychosis-like behaviors in mice, he disabled it by removing GSK3β in D2R-containing neurons. This interfered with apomorphine-induced rearing as well as amphetamine-induced hyperlocomotion and prepulse inhibition, but not cognitive tasks. Removing GSK3β's target, β-catenin, from D1R-containing neurons induced more amphetamine-induced psychosis, whereas removing it from D2R-containing neurons induced less. Together, the results implicate this pathway in psychosis-like behaviors and antipsychotic response, and the extent of its contribution may be further delineated with D2R mutants engineered to selectively signal through either the G protein or the β-arrestin-2 pathway.

The schizophrenia risk gene CAV1 is both propsychotic and required for antipsychotic drug activity at 5-HT2A serotonin receptors in vivo

In a later talk (see below), John Allen (University of North Carolina) reported new efforts to discover D2R ligands that selectively activate the β-arrestin-2

Table 2. Translational neuroscience approaches to schizophrenia

	Advantage	Disadvantage
Olfactory neurons	• Live immature neuron-like cells • Sufficiently homogeneous • *In vitro* functional study • Easy for preparation	• May not completely represent brain tissue • May not be able to chase several developmental phases
iPS cells—derived neurons	• Live neurons • *In vitro* functional study • May be able to differentiate into many lineages, several developmental stages	• Laborious, high cost • Heterogeneity • Reprogramming artifacts
iN cells (induced neuronal cells)	• Live neurons • *In vitro* functional study • Potentially high-throughput	• Differentiation into many lineages has not yet been fully established
Autopsied brains	• Brain tissue	• Many confounding factors • Functional assay not available • Less of a link to developmental process

pathway but not the G protein pathway. In a chemical biology effort in Bryan Roth's laboratory (University of North Carolina), the group screened hundreds of new compounds and came up with two that are biased for their signaling activity at the D2R-β-arrestin-2 pathway. When administered to mice, these arrestin-biased compounds decreased PCP- or amphetamine-induced hyperlocomotion without increasing catalepsy. These results suggest that selective D2R-β-arrestin-2 signaling and recruitment may contribute to antipsychotic activity without motor side effects.

The dopamine D1-D2 receptor heteromer: novel signaling complex with potential role in schizophrenia

In another twist for dopamine signaling, Susan George (University of Toronto) presented her evidence for a novel dopamine receptor made up of a D1R and a D2R. This combination receptor offers a new mode of signaling for dopamine because it activates a G protein pathway that is not induced when either receptor is activated alone. The D1-D2 heteromers occur endogenously in the brain and are enriched in the nucleus accumbens and globus pallidus of rats. Using a competitive binding assay to detect a high-affinity state for a D2 agonist in the D1-D2 heteromer, George found this state is enhanced

in rats treated with amphetamine and in the globus pallidus of the postmortem brain in schizophrenia. The work suggests both that abnormal coupling between D1 and D2 could be a molecular marker for pathological states in the brain and that this coupling may be a new treatment target—something that has already been explored in mouse models of depression.

Serotonin signaling is also thought to contribute to schizophrenia symptoms, and the atypical antipsychotics have been designed to also target 5-HT$_{2A}$ serotonin receptors, following suggestions that this activity set clozapine apart from the typical antipsychotics. Prompted by a rare CNV found in the caveolin-1 (*CAV1*) gene in a case of schizophrenia,[18] John Allen explored the state of serotonin signaling in knockout mice missing *CAV1*. *CAV1* encodes a scaffolding protein involved in clustering diverse signaling molecules together, including 5-HT$_{2A}$ receptors.[19] Loss of *CAV1* resulted in an increased sensitivity to PCP-induced hyperlocomotion and disrupted prepulse inhibition, suggesting a propsychotic-like effect in the mice. Loss of the *CAV1* gene also attenuated the activity of atypical antipsychotics and 5-HT$_{2A}$ antagonists in behavioral studies (Fig. 4). Similarly, these mice made fewer head twitches in response to a hallucinogenic 5-HT$_{2A}$ agonist and also disrupted 5-HT$_{2A}$ signaling

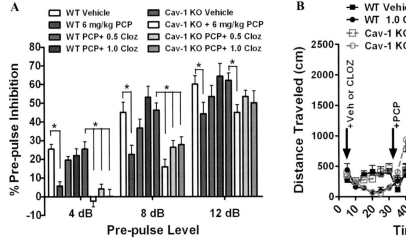

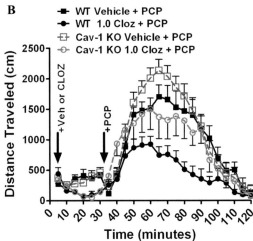

Figure 4. Genetic deletion of schizophrenia risk gene caveolin-1 in mice annenuates antipsychotic drug activity. (A) Littermate wildtype (WT) and caveolin-1 knockout (Cav-1 KO) mice were injected with vehicle, 0.5, or 1.0 mg/kg clozapine (CLOZ) and were followed 30 min later by 6 mg/kg phencyclidine (PCP) and immediately placed into acoustic startle response chambers. Clozapine-normalized PCP-disrupted prepulse inhibition (PPI) in WT but not in Cav-1 KO mice at the 4 dB and 8 dB prepulse levels (mean ± SEM, $n = 12$ littermate pairs; *, $P < 0.05$ versus WT Saline or Cav-1 KO Saline at each prepulse level). (B) Littermate WT and Cav-1 KO mice received vehicle or 1.0 mg/kg CLOZ and were placed into open field locomotion chambers and allowed to acclimate; 30 min later, mice were injected with PCP and the total distance traveled (cm) was determined. Clozapine was less effective at suppressing hyperlocomotion in Cav-1 KO mice (mean ± SEM, $n = 12$ littermate pairs).[20]

in cortical neurons. The number of 5-HT$_{2A}$ receptors was normal in these mice, which suggests that loss of the *CAV1* scaffold mislocalized 5-HT$_{2A}$ receptors and their downstream effector molecules, compromising their function and resulting in compromised antipsychotic activity.[20]

Although the crowd was dwindling, energy remained high at the end of the talks, with several people saying that they felt optimistic about the prospects for discovering truly innovative drugs for schizophrenia. One participant raised the issue of specificity, asking how disturbances to intracellular signaling pathways available to all cells can produce the malfunctions in specific brain circuits observed in schizophrenia. Moghaddam suggested that patterns of metabolic activity, combined with aberrant signaling, somehow target certain circuits for dysfunction. John Krystal (Yale University) suggested a genetic explanation, noting that susceptible brain regions in schizophrenia are the most recently evolved and thus could be the most genetically labile. Either way, a challenge will be to deliver treatment to ailing brain circuits without disrupting those that are functioning normally.

In his closing remarks, Krystal called the meeting "a next-generation conference" with research beginning to meet the urgent need for mechanistically

novel compounds. "What's exciting is not how far we've come, but the possibility that we might be getting to the point where we can use science to guide psychiatry.

References

1. Sullivan, P.F. Schizophrenia Psychiatric Genome-Wide Association Study Consortium: genome-wide association study of schizophrenia identifies five novel loci. Submitted
2. Sullivan, P.F. 2010. The Psychiatric GWAS Consortium: big science comes to psychiatry. *Neuron* **68:** 182–186.
3. Nestler, E.J. & S.E. Hyman. 2010. Animal models of neuropsychiatric disorders. *Nature Neurosci.* **13:** 1161–1169.
4. Huffaker, S.J., J. Chen, K.K. Nicodemus, *et al.* 2009. A primate-specific, brain isoform of KCNH2 affects cortical physiology, cognition, neuronal repolarization and risk of schizophrenia. *Nature Med.* **15:** 509–518.
5. Guidotti, A., J. Auta, Y. Chen, *et al.* 2011. Epigenetic GABAergic targets in schizophrenia and bipolar disorder. *Neuropharmacology* **60:** 1007–1016.
6. Guidotti, A., E. Dong, M. Kundakovic, *et al.* 2009. Characterization of the action of antipsychotic subtypes on valproate-induced chromatin remodeling. *Trends. Pharmacol. Sci.* **30:** 55–60.
7. Dong, E., D.R. Grayson, A. Guidotti & E. Costa. 2009. Antipsychotic subtypes can be characterized by differences in their ability to modify GABAergic promoter methylation. *Epigenomics* **1:** 201–211.
8. Cheung, I., H. P. Shulha, Y. Jiang, *et al.* 2010. Developmental regulation and individual differences of neuronal H3K4me3

epigenomes in the prefrontal cortex. *Proc. Natl. Acad. Sci. USA* **107:** 8824–8829.

9. Peter, C.J. & S. Akbarian. 2011. Balancing histone methylation activities in psychiatric disorders. *Trends. Mol. Med.* **17:** 372–379.

10. Conn, P.J., C.W. Lindsley & C.K. Jones. 2009. Activation of metabotropic glutamate receptors as a novel approach for the treatment of schizophrenia. *Trends Pharmacol. Sci.* **30:** 25–31.

11. Ayala, J.E., Y. Chen, J.L. Banko, *et al.* 2009. mGluR5 positive allosteric modulators facilitate both hippocampal LTP and LTD and enhance spatial learning. *Neuropsychopharmacol.* **34:** 2057–71.

12. Mao, Y., X. Ge, C.L. Frank, *et al.* 2009. Disrupted in schizophrenia 1 regulates neuronal progenitor proliferation via modulation of GSK3beta/beta-catenin signaling. Cell **136:** 1017–1031.

13. Beaulieu, J.M., R.R. Gainetdinov, & M.G. Caron. 2009. Akt/GSK3 signaling in the action of psychotropic drugs. *Annu. Rev. Pharmacol. Toxicol.* **49:** 327–247.

14. Pan, J.Q., M.C. Lewis, J.K. Ketterman, *et al.* 2011. AKT kinase activity is required for lithium to modulate mood-related behaviors in mice. *Neuropsychopharmacol.* **36:** 1397–1411.

15. Sawa, A. & N.G. Cascella. 2009. Peripheral olfactory system for clinical and basic psychiatry: a promising entry point to the mystery of brain mechanism and biomarker identification in schizophrenia. *Am. J. Psychiatry* **166:** 137–139.

16. Takahashi, K., K. Tanabe, M. Ohnuki, *et al.* 2007. Induction of pluripotent stem cells from adult human fibroblasts by defined factors. *Cell* **131:** 861–872.

17. Pang, Z.P., N. Yang, T. Vierbuchen, *et al.* Induction of human neuronal cells by defined transcription factors. *Nature* **476:** 220–223.

18. Walsh, T., J.M. McClellan, S.E. McCarthy, *et al.* 2008. Rare structural variants disrupt multiple genes in neurodevelopmental pathways in schizophrenia. *Science* **320:** 539–543.

19. Allen, J. A., R. A. Halverson-Tamboli & M. M. Rasenick. 2007. Lipid raft microdomains and neurotransmitter signalling. *Nat. Rev. Neurosci.* **8:** 128–140.

20. Allen, J.A., Y. P., V. Setola, M. Farrell, & B.L. Roth. 2011. Schizophrenia risk gene CAV1 is both pro-psychotic and required for atypical antipsychotic drug actions *in vivo. Transl. Psychiatr.* e33; doi:10.1038/tp.2011.35. Published online 16 August 2011.